The Boundless Universe

The Boundless Universe

Astronomy in the New Age of Discovery

Sidney C. Wolff

Rio Nuevo Publishers
Tucson, Arizona

Rio Nuevo Publishers®
P. O. Box 5250
Tucson, AZ 85703-0250
(520) 623-9558, www.rionuevo.com

Library of Congress Cataloging-in-Publication Data

Wolff, Sidney C.
The boundless universe : astronomy in the new age of discovery / Sidney C. Wolff.
pages cm
Includes index.
ISBN 978-1-940322-09-4 (pbk.)—ISBN 1-940322-09-X (pbk.)
1. Astronomy—Popular works. 2. Astronomical observatories—Popular works. 3. Scientific satellites—Popular works. I. Title.
QB44.3.W65 2016
520—dc23

2015030555

Book design: Preston Thomas, Cadence Design Studio
Managing Editor: Aaron Downey

Front cover image: The Milky Way rising over the Frank N. Bash Visitors Center at McDonald Observatory. This amphitheater is used for star parties. © Ethan Tweedie Photography

Title page image: Milky Way over the Alabama Hills, near Lone Pine, California. © Steve Rengers Photography

Back cover image: The Rosette Nebula. This is a region where stars are forming. Winds of particles from hot, blue, young stars have cleared gas away from the center. This is a false color image. Glowing hydrogen gas is shown in red, oxygen emission is shown in green, and sulfur emission is shown in blue. The image was taken at Kitt Peak.
T. A. Rector/University of Alaska Anchorage, WIYN, and NOAO/AURA/NSF

Printed in Korea.

10 9 8 7 6 5 4 3 2 1

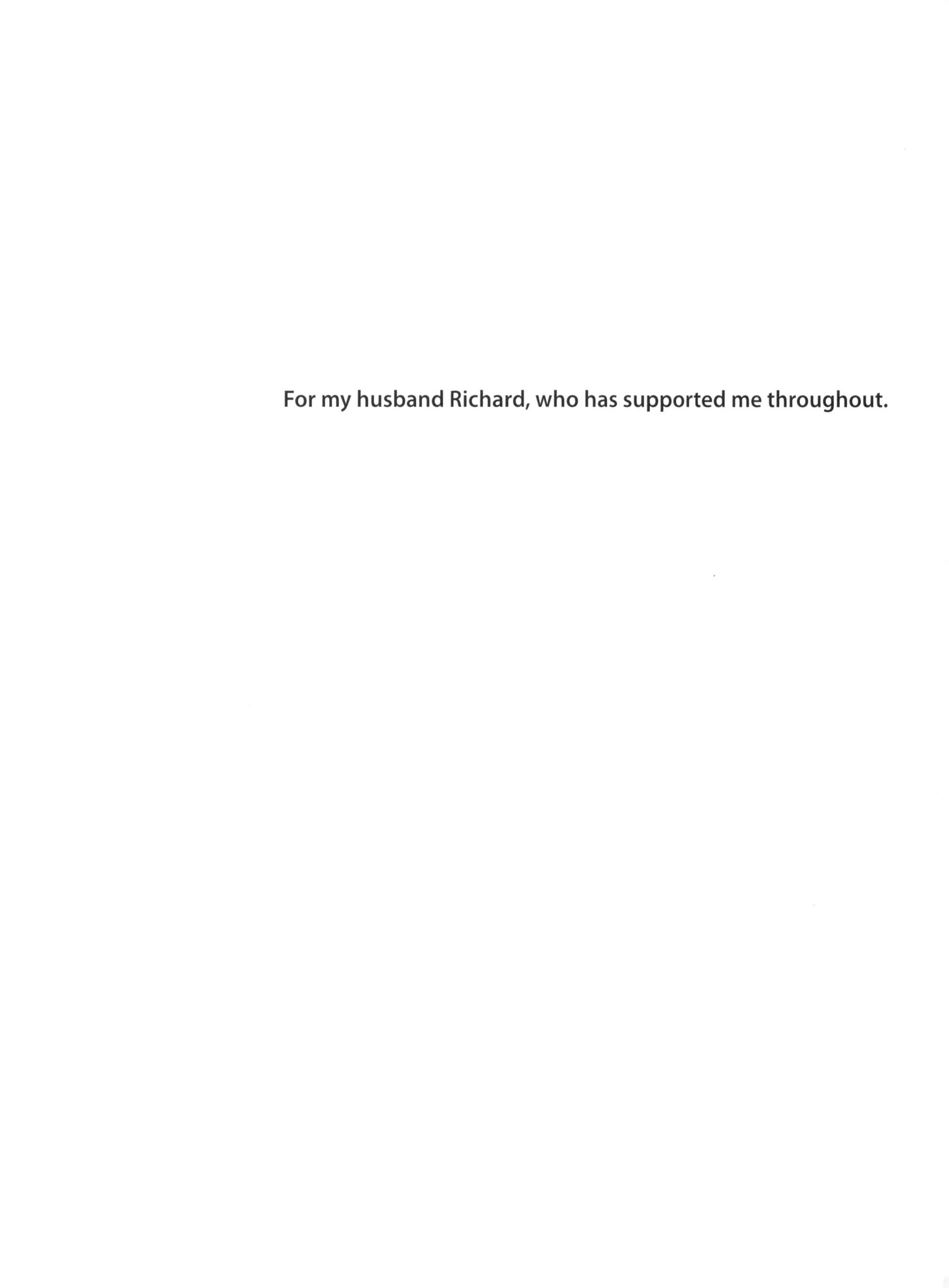

For my husband Richard, who has supported me throughout.

Contents

Introduction

How old is the Universe? What is it made of? How has it evolved? When did the Earth form and what will be its fate? Is there life elsewhere in the solar system or on distant planets orbiting other stars?

Look up the Age of Exploration on the Web, and you will find a description of adventurous sea voyages that began in the 1400s. European rulers were anxious to discover a reliable sea route to Asia. Their surprising discoveries in North and South America led to radically changed maps of the world.

Now we live in a new Age of Exploration, but in this case the Universe is our destination, and telescopes provide our ships of discovery. The first three chapters of this book describe the major ground-based optical observatories in the continental United States, all of which are located in the Southwest, from California to Texas; explain why there are now newer observatories in Hawaii and Chile; and summarize technical innovations that make it possible to build a new generation of truly giant telescopes.

To continue the analogy, the final three chapters are dispatches from the frontier. Scientists are combining observations from ground-based observatories with measurements made from space to map the Universe and describe its contents. Spacecraft have now visited every planet in our solar system. Thousands of planets around other stars have been discovered. We have observed galaxies as they were when the Universe was only 5 percent as old as it is now. We can see stars and planets in the earliest stages of formation. And we are just beginning the serious search for life elsewhere—life that is possibly in our own solar system but more likely on planets orbiting distant stars.

In the first Age of Exploration, reports about the New World had to rely on words and perhaps some drawings. The amazing images from ground-based telescopes, the Hubble Space Telescope, and planetary missions can actually show us what the Universe is like with an accuracy and precision that drawings and words cannot match.

The many images in this book will take you on a virtual journey to major ground-based observatories, then through our solar system, and on to the most distant objects observed so far in the Universe. But by no means are we at the end of our current Age of Exploration. New telescopes, currently under construction on the ground and in space, will continue to make new discoveries. This book describes some of the plans for future research, and astronomers are committed to making their new results publicly available and understandable. This book makes a number of suggestions about how to continue to learn about new discoveries. It also describes how you can participate directly through citizen science projects that help scientists with their research. Join the adventure!

Messier 16, the Eagle Nebula. This image was taken with a 36-inch telescope, which is one of the smaller telescopes on Kitt Peak. Located in the constellation of Serpens, the Serpent, the Eagle Nebula is a very luminous open cluster of stars surrounded by dust and gas. This is a region of active star formation. This image was created by combining three separate images that recorded emission from hydrogen (green), oxygen (blue), and sulfur (red).

T. A. Rector (NRAO/AUI/NSF and NOAO/AURA/NSF) and B. A. Wolpa (NOAO/AURA/NSF)

1 Six Observatories of the Southwest

What do a land baron in California, a Boston Brahmin, a steel magnate, an oil tycoon, and a banker all have in common? Each decided to invest some of his wealth to develop a new astronomical observatory in the southwestern United States. Then in the 1950s, the U.S. National Science Foundation created a national observatory that would be open to all astronomers. Through these investments, the United States became the world leader in astronomical research for more than a century. This chapter describes some of the colorful history of the founding of these six observatories. All of them continue to do important research today.

Kitt Peak National Observatory

Near Tucson, Arizona

On May 7, 1955, the young astronomer Helmut Abt climbed into a two-seater Cessna to survey mountains in the Southwest in order to find a site for a new national astronomy observatory. Up until that time, all observatories in the United States were privately funded, and only astronomers at the universities that operated the telescopes could use them. For convenience, most universities located their observatories on or near their campuses—whether or not the sites were good for astronomy. Abt's charge was to find the very best site in the United States for the national observatory.

Flying was more of an adventure in the 1950s than it is now. Abt's pilot John Casparis was a Texan who had been flying since before Abt was born. The stories Casparis told about his 33,000 hours of flying experience dated back to World War I. More recently, he said, he made part of his living from the bounty offered in sheep-herding country for killing eagles. His technique was to fly very near the cliffs where the eagles lived, shoot them from his plane, remember where they fell, and then go back later to pick up the carcasses

The first primitive road up Kitt Peak. The vehicle is hauling one of the last segments of the tower that was used to measure the quality of astronomical images (seeing) on Kitt Peak. The road today is much improved!

John Glaspey/NOAO/AURA/NSF

in order to collect the bounty. In order to land at night at his base in Marfa, Texas, Casparis flew his plane low over the town and turned his engine off and on several times to alert his friends to go to the local airstrip and turn on the landing lights.

Despite the pilot's experience, Abt had some tense moments. After exploring southern New Mexico, Casparis landed in Bisbee to refuel, but there was no gas. Fortunately, he had enough fuel to reach Nogales, which did have gas but no food for lunch. Abt and Casparis had to settle for 5-cent coffee and a candy bar from vending machines. After three days of flying over mountains and forests where there were no safe landing sites, the four-cylinder engine began to misfire. Again, fortunately, this didn't happen until the return trip to Texas, where Casparis was able to make an emergency landing on a deserted airstrip.

From an astronomical standpoint, the trip was a complete success. In three days, Abt traveled 2,000 miles and identified 150 potential sites. The prime California site was ultimately eliminated because there were already three major observatories in that state, and astronomers wanted to take advantage of Arizona's complementary weather patterns. California has the highest percentage of clear weather in the summers, when nights are short. In contrast, Arizona has the highest percentage of cloudy weather during the summer monsoon season and a large percentage of clear nights in the winter, when the nights are long.

Trips by Jeep and horseback narrowed the choices further to several sites in northern Arizona and to Kitt Peak outside of Tucson. On their first Jeep trip up Kitt Peak, Abt and Aden Meinel, who was in charge of the effort to select the final site and who became the first director of Kitt Peak National Observatory, were unable to reach the summit because the manzanita was too thick. Later reconnaissance trips were made on horseback with Tohono O'odham guides.

The next challenge was obtaining permission to install site-testing equipment. All of the potential sites except Kitt Peak were on land controlled by the U.S. Forest Service, and permission was easily obtained. Kitt Peak, however, is located on Tohono O'odham land, and approval of the Schuk Toak District Council was required in order to put in a primitive road and install observing equipment. On October 28, 1955, members of the Tribal and District Councils were invited to Steward Observatory at the University of Arizona to see an operating observatory, learn about astronomy, and view the Moon. The Council members were very excited about what they saw, and they were particularly impressed by the contrast in the appearance of the Moon with and without a telescope. After this visit, the District Council agreed to make Kitt Peak available to the astronomers, whom they called the people "with long eyes." There was no word in the Tohono O'odham language for "astronomer."

The initial intention was to conduct a thorough site survey of Kitt Peak and several other candidate mountains, with particular attention to measuring the "seeing." Astronomers use the term *seeing* to refer to the blurring and twinkling of stars and other

Telescopes on Kitt Peak. The tallest dome is the home of the Mayall 4-meter telescope. The smaller domes in front of it are operated by the University of Arizona. Isolated storms are often seen over the desert during the summer monsoons, but the skies usually clear after midnight. Darryl Willmarth/NOAO/AURA/NSF

astronomical objects by turbulence in the Earth's atmosphere. The technical difficulty of measuring seeing with the requisite accuracy, combined with budget limitations and the desire to select a site quickly, made it impossible to obtain quantitative results for multiple sites. In 1960, astronomers concluded that weather data and half a year's worth of automatically recorded seeing data were enough to show that Kitt Peak was a good astronomical site. Limited weather data and visual estimates of seeing at the other sites indicated that they were probably inferior. Kitt Peak thus became the site of the first national optical astronomy observatory. Also in 1960, Helmut Abt became one of the first astronomers to join the scientific staff of the fledgling observatory, and he has remained there throughout his entire career.

There are several reasons that a national observatory was created in the 1950s. First was the establishment of the National Science Foundation (NSF) in 1950 by the U.S. Congress. The mission of the NSF was to ensure continuation of the scientific and technological progress that played such a critical role during World War II by supporting basic research in the United States. An initial NSF project was to establish an observatory that would provide every astronomer, independent of institutional affiliation, access to state-of-the-art telescopes. The second impetus for the national observatory was the increasing ease of plane travel, which meant that observatories no longer needed to be located close to the astronomers who would use them. Kitt Peak could be reached with one day of travel from anywhere in the United States. Astronomers could go to the observatory, use the telescopes for a few nights, and then return to teach at their home universities. And the third reason was the availability of new electronic sensors that enabled the measurement of the brightness of stars with an accuracy of one percent or so—but only if skies were completely clear. This put a premium on building observatories in the U.S. region with the largest percentage of cloud-free nights: the American Southwest.

Kitt Peak Today

Kitt Peak is home to a large number of optical, solar, and radio telescopes operated by Kitt Peak National Observatory (KPNO) and several university groups. The two largest telescopes managed by KPNO are the 4-meter (157-inch) Mayall and the 3.5-meter (138-inch) WIYN (originally built by the Universities of Wisconsin and Indiana, Yale, and the National Optical Astronomy Observatory, of which KPNO is a part). Both are general-purpose telescopes equipped with multiple instruments for imaging and spectroscopy of stars and galaxies. Beginning in 2018, most of the observing time on these two telescopes will be devoted to large survey programs that will contribute to two of the hottest topics in modern astronomy: What is dark energy? And are there potentially habitable planets around other stars?

The Mayall Telescope will be used to study dark energy. We know that the expansion of the Universe is speeding up as it ages. The goal of the Mayall observations is to measure how the rate of expansion changes with the aging of the Universe and hence to characterize *dark energy*, the mysterious cause of the acceleration (see Chapter 6). The observations will be made with the Dark Energy Spectroscopic Instrument (DESI), which is being funded by the Department of Energy. DESI will obtain optical spectra for tens of millions of galaxies in order to create a 3-D map of the galaxies and their velocities out to a distance of 10 billion light-years.

The new spectroscopic instrument for the WIYN Telescope will be designed to measure the subtle changes in velocity of a star caused by a planet in orbit around it (see Chapter 5). This project is a key component of the NASA-NSF Exoplanet Observational Research (NN-EXPLORE) program. These velocity measurements will be used to confirm that candidate exoplanets discovered by NASA spacecraft are actually planets and not low-mass stars; to derive the mass of the planets; and to determine whether the planets are rocky like the Earth and potentially habitable, or instead are balls of gas like Jupiter.

For information on visiting Kitt Peak National Observatory, go to www.noao.edu/kpvc.

Lick Observatory

Near San Jose, California

In 1846, James Lick decided to move to California, which the United States had just seized from Mexico. Lick had already had an adventurous life. He was born in 1796 in Pennsylvania and was trained by his father as a cabinetmaker. After stops in Buenos Aires and Valparaiso, Lick moved to Lima, Peru, in 1836. There he specialized in making cases for pianos and accumulated a small fortune.

Lick arrived in California on January 7, 1848, just 17 days before the discovery of gold at Sutter's Mill. He brought with him $30,000 in gold coins and 600 pounds of chocolate made by his neighbor Domenico Ghirardelli, which he sold at a profit. (At Lick's urging, Ghirardelli moved to California in 1849, and in 1852 he launched a business that became the Ghirardelli Chocolate Company.)

When Lick arrived, San Francisco had only 800 inhabitants. Because it was the port closest to the gold rush mines, Lick knew that the city would grow rapidly. He began to buy land there and also near San Jose, which served briefly as the state capitol. As a result of his investments in land, Lick became extremely wealthy.

In 1873, Lick suffered a severe stroke. He then began to think about how he might use his wealth to establish a monument to himself. Options included large statues of himself and his parents and a pyramid, which would be larger than the great pyramid in Egypt, on Market Street in San Francisco. Through meetings with several leading astronomers, Lick had long been fascinated by astronomy, and in the end he decided to invest a large portion of his assets in building an observatory. His first choice for a site was Market Street, where the observatory could be easily seen and admired. His astronomer friend George Davidson, however, quickly persuaded him that the haze and smoke in the city would limit the science that could be done with a new telescope.

Most existing observatories were located at low altitude. Davidson knew, however, from his own measurements with a portable telescope, that astronomical images were sharper when viewed from mountaintops rather than at sea level. Building an observatory on top of a mountain would be unprecedented, but Davidson persuaded Lick he should pioneer the development of the first large mountaintop observatory in the United States. Lick's initial choice of a site was property he owned near Lake Tahoe, but access during the winter proved to be too difficult. In 1876, just a few months before he died, Lick agreed that his observatory should be built on Mount Hamilton, whose summit at an altitude of 4,400 feet was visible from his property in San Jose.

Construction of the observatory became the responsibility of the trustees of Lick's estate, and completion took another decade. On January 8, 1887, Lick's coffin was transported to

The open dome of the 3-meter (120-inch) Shane Telescope at Lick Observatory. The dome at the far left on the horizon houses the 36-inch Lick refractor. © Laurie Hatch

Mount Hamilton and placed in the foundation of the telescope that had been built in his name. The 36-inch refracting telescope (see Chapter 3 for a description of refracting and reflecting telescopes), the largest of its type at that time, was completed in 1888, and then title for the observatory was transferred to the University of California, which continues to operate it today.

Lick Observatory Today

The largest telescope on Mount Hamilton today is the 120-inch Shane reflector, which was commissioned in 1959. The 120-inch-diameter primary mirror was originally made as a test by Corning prior to casting the 200-inch Palomar mirror. Lick Observatory is strongly affected by light pollution from San Jose but continues to do important research.

The newest telescope on the mountain is the Automated Planet Finder. This 2.4-meter reflecting telescope operates completely robotically. That means that the telescope has been programmed to operate throughout the night with no human intervention. Each night its systems check the weather, decide which stars to observe, and move the telescope from one star to the next until dawn. The measurements from this telescope are being used to search for Earth-sized planets around nearby stars (see Chapter 5).

A second robotic telescope is used to discover supernovae, which are spectacular stellar explosions that can briefly outshine an entire galaxy before fading from view (see Chapter 6). The observatory is also used for training students and testing new technologies before they are deployed on the 10-meter Keck telescopes in Hawaii (see Chapter 3).

For information on visiting Lick Observatory, go to www.ucolick.org/main/index.html.

Lowell Observatory

Flagstaff, Arizona

Percival Lowell, unlike James Lick, decided to build an observatory because he wished to use it himself, especially to observe the planet Mars. The Lowells first arrived in Boston in 1639. Because of their many intellectual contributions and achievements in business in the 19th and 20th centuries, the Lowells were widely recognized as one of America's most distinguished families. Percival Lowell graduated from Harvard in 1876 with distinction in mathematics. His thesis, a harbinger of things to come, was on the "Nebular Hypothesis," which described then-current ideas about how the solar system might have formed.

In 1893, Lowell decided to devote himself full time to the study of astronomy, and most particularly to observations of Mars. An Italian expert on Mars, Giovanni Schiaparelli, had retired in 1892 because of failing eyesight. What intrigued Lowell was Schiaparelli's claims that he had seen narrow lines on the surface of Mars. In Italian, he called these lines *canali*, which means grooves, but the English translation became canals. Was this evidence of life on Mars?

In order to study faint markings on the surface of Mars, Lowell knew he needed a site that was very dark—distant from contamination by city lights and smog. He also needed very sharp images, which meant his observatory had to be on a mountain. The challenge was to find such a site very quickly because Mars was to reach one of its closest points to Earth in the summer of 1894. Lowell hired A. E. Douglass, who had performed site surveys in South America for Harvard, to search for a suitable location in the Arizona Territory. Arizona at that time was still the Wild West. Its total population was less than 100,000. It did not become a state until 1912. And Geronimo had surrendered for the final time in only 1886.

Douglass began his search in Bisbee in early March of 1894, and briefly examined sites near Tucson, Phoenix, and Prescott before reaching Flagstaff in early April. It was here that he had the first good weather of his trip. The city offered him some land, which became known as Mars Hill, for the planned observatory. With a dome and a borrowed 18-inch refractor from Harvard, the observatory was up and running by the end of May—an amazing feat! Lowell, Douglass, and Harvard astronomer Edward Pickering made observations of Mars nearly every night from then until mid-December, and nearly 1,000 drawings of its surface were produced (photography was not yet sensitive enough). A 24-inch refractor was installed at the observatory in 1897. The lens was made by Alvan Clark and Sons, the same company that made the 36-inch lens for the Lick refractor.

Questions remained, however, about whether the markings on Mars reported by Lowell and others were real or an optical illusion. In 1901, Douglass set up an experiment; he placed small globes on the ground at appropriate distances and put markings on them. Lowell agreed to observe the globes through a telescope. For one, he drew two lines where in fact there was only one. This and other tests showed that visual observations that claimed to see narrow linear markings on Mars were unreliable. Three months later, Douglass lost his job at Lowell. In 1906, he joined the University of Arizona in Tucson, where he subsequently became head of the Steward Observatory. There he also pursued research on dendrochronology, which is the science of using growth rings in trees to study long-term trends in climate and to age-date artifacts found at archaeological excavations. The Laboratory for Tree Ring Research, which Douglass founded in 1937 at the University of Arizona, is recognized worldwide as a leader in tree-ring research.

Lowell Observatory Today

Two historic telescopes are located on Mars Hill in Flagstaff—the Clark refractor, which was used by Percival Lowell, and the Pluto telescope, so named because it was with this telescope that Clyde Tombaugh discovered the (dwarf) planet Pluto. After Tombaugh discovered the new planet, the Lowell Observatory staff offered people worldwide the opportunity to suggest a name. Venetia Burney, an 11-year-old schoolgirl in England, suggested the name Pluto. In Greek mythology, Pluto, the god of the underworld, stays hidden in a dark, remote place—just like the planet Pluto. Tombaugh went on to discover

The 24-inch Clark refractor at Lowell Observatory. This telescope was used by Percival Lowell and others to observe Mars. Today, visitors to Lowell have an opportunity to look through this historic telescope. Tom Alexander/Lowell Observatory

EXPLORING ON YOUR OWN

Visiting Observatories

All of the observatories described in this chapter offer programs for public visitors. Visitor centers provide information about each observatory and include displays about astronomy. Some observatories offer guided tours, while others are self-guided. In either case, the largest telescope(s) can usually be viewed from a glassed-in visitors' gallery during the daytime. At night, the telescopes are used for research. Astronomers are able to apply for time once or twice each year. Usually, half to three-quarters of the proposals have to be rejected because there are simply not enough nights available to meet demand. The astronomers whose proposals are judged to be the best are awarded a few nights each, typically around three nights, but sometimes only half a night. If the weather is cloudy, then the astronomer must submit a new application. There are no rain checks in astronomy!

Most of these observatories are in very scenic locations—an added bonus for the visitor. Some observatories offer opportunities to look through a smaller telescope at night; reservations are usually required. Before making a visit, be sure to check the observatory website for hours. Detailed information can be found by typing the name for each observatory into your favorite Internet search engine.

The University of Arizona offers public lectures in the evening and also operates a nightly observing program on Mount Lemmon. Check your nearby university to see whether it has any similar offerings.

▲ **The Very Large Array (VLA), a radio interferometer in New Mexico.** NRAO/AUI and NRAO

Scattered throughout the Southwest are bed and breakfast inns that have a small observatory nearby. Use a search engine with the phrase "astronomy bed and breakfast" to find locations and descriptions of the observing opportunities.

This book focuses on optical observatories, but one of the world's largest radio astronomy observatories is also located in the Southwest and is open to visitors. The Very Large Array (VLA) is located in a radio quiet zone about 60 miles west of Socorro, New Mexico. The VLA consists of 27 radio antennas mounted on rails so that they can be moved. Each antenna is 25 meters (82 feet) in diameter. The antennas are spread out into a Y-shaped configuration that can be as much as 22 miles across.

The VLA telescopes together form an interferometer. This may seem like a strange name because the telescopes do not interfere with each other but work cooperatively to see fine detail in astronomical objects, such as galaxies. Scientists call the ability to see fine detail resolution. The resolution of any telescope depends on its diameter. If you look at the Moon with a small telescope you can see more detail in the craters and mountain ranges than you can if you look with just your eye. That is because the lens or mirror in the telescope is larger than the pupil of your eye, and so the resolution of the telescope is higher.

By combining the light from, say, two telescopes that are a mile apart, radio astronomers can create a virtual telescope that offers the same resolution as a telescope that is a mile in diameter. It is much cheaper to build two smaller telescopes separated by a mile than to build a telescope that is a mile in diameter. The VLA can provide a resolution equivalent to a telescope that is 22 miles in diameter!

For information on visiting the Very Large Array, go to www.vla.nrao.edu. ●

DISCOVERY OF THE PLANET PLUTO

January 23, 1930

January 29, 1930

These are copies of small sections of the discovery plates showing images (those marked) of Lowell's mathematically predicted trans-Neptunian planet afterward named PLUTO. It was found by Mr. C. W. Tombaugh on February 18, 1930, while engaged in the search program and upon examination of these plates.

Lowell Observatory Photograph

Top: Clyde Tombaugh using a blink comparator to search for a new planet. This device allows the user to blink quickly between two images taken at different times and look for changes. A planet would be expected to move and be in different locations on the two plates. Stars and galaxies, of course, don't move perceptibly on short time scales. Think of the patience required to compare the hundreds of thousands of stars on the many pairs of plates that Tombaugh examined. Lowell Observatory Archive

Bottom: The discovery plates. Lowell Observatory Archive

many asteroids. A small portion of his ashes is aboard the New Horizons spacecraft that made the first close-up reconnaissance of distant Pluto in 2015 (see Chapter 4).

The premiere facility of Lowell Observatory is the Discovery Channel Telescope (DCT), which began scientific operations in 2015 (see Chapter 3). This telescope is located on a dark site about 50 miles southeast of Mars Hill. Construction was funded by a major gift from the Discovery Channel and is operated in partnership with Boston University, the University of Maryland, the University of Toledo, and Northern Arizona University. The telescope is equipped with several instruments, including a 36-megapixel camera.

For information on visiting Lowell Observatory, go to http://lowell.edu.

Mount Wilson and Palomar Observatories

Los Angeles, California and Palomar Mountain, California

George Ellery Hale was the preeminent telescope builder of the first half of the twentieth century. Four different times, he initiated the effort to build the largest telescope in the world. The first of these was the 40-inch refractor at Yerkes Observatory, which was finished in 1897. In 1890, Hale learned that the University of Southern California had abandoned plans to build a large refracting telescope, but the two glass lenses for the project had already been manufactured. Hale was then on the faculty of the newly established University of Chicago, and he persuaded its President, William Rainey Harper, that the university should acquire the glass disks and build an observatory. With financing from Charles Yerkes, who made his fortune by building public transportation in Chicago, the observatory was built on the shores of Lake Geneva in southern Wisconsin.

Even before the Yerkes 40-inch was completed, Hale was already planning to use new technology to build an even bigger telescope at a site with fewer cloudy nights. This facility would be necessary for *astrophysics*—the application of the principles of physics to the study of astronomical objects.

Throughout most of the nineteenth century, *astronomy* involved primarily measurements of the positions and motions of stars. This information could be used to derive the distances to stars and, in the case of binary stars, to estimate their masses. In 1863, however, William Huggins obtained the first photograph of the spectrum of a star. He spread the white light from the star out into its rainbow of colors from blue to red (see Chapter 5 for more discussion of this technique, which is called spectroscopy) and discovered that the stars are made of the same kinds of elements, including for example, hydrogen, sodium, and iron, as are found here on Earth. For the first time, astronomy could be combined with what was already known from studies in physics laboratories to understand celestial objects. The field of astrophysics was born.

Hale was trained as a solar astronomer, and he had already used spectroscopy to study the atmosphere of the Sun. Stars, however, are much fainter than the Sun, and Hale knew that large telescopes would be required to obtain spectra of any but the very brightest stars. He also knew that telescopes based on lenses could not be significantly larger than the Lick 36-inch and Yerkes 40-inch refractors (see Chapter 3). In 1896, Hale's father had bought him a 60-inch mirror that could serve as the key optical component in a still-larger telescope—but one that used reflecting mirrors rather than lenses.

Hale began dreaming of California. In 1889, Harvard astronomer Edward Pickering had already made observations from Mount Wilson, just outside of Los Angeles. Although he found that the observing conditions were much better than on the east coast, Pickering concluded that Mount Wilson was unsuitable for astronomy because of its primitive living conditions, lack of a suitable water supply, and too many rattlesnakes. Hale felt that the superior observing conditions trumped these logistical difficulties and began to search for funding to develop a solar and nighttime observatory on Mount Wilson.

At just the right time, Andrew Carnegie supplied an opportunity. In 1902, he founded the Carnegie Institution of Washington as an organization to enable scientific discovery. Shortly afterward, Hale was invited by his old friend Pickering to serve on the Advisory Committee on Astronomy that had been established by the Carnegie Institution. Hale later said that, "The provision of a large endowment solely for scientific research seemed almost too good to be true."

In June 1903, Hale paid his first visit to Mount Wilson. Transportation up the mountain was by burro. In his diary, Hale recorded the costs: unbroken burro, $15; broken to trail, $35; cost of food, 15 cents a day. Observations during the remainder of that year showed that Mount Wilson is indeed an excellent observing site for both solar and stellar astronomy.

After submitting plans to the Carnegie Institution, Hale received notification in December 1904 that he had been granted sufficient funds to build both a solar and a 60-inch nighttime telescope on Mount Wilson. On December 20, 1908, the first images were taken with the 60-inch telescope. Hale reported that the results were "admirable," with small perfect images of stars and well-defined details in the Orion Nebula. For a decade, the 60-inch was the largest telescope in the world.

In 1906, *two years before the 60-inch was completed,* Hale began talking about the possibility of building a 100-inch telescope on Mount Wilson. This would be a tremendous technical challenge because no piece of glass this large had yet been cast. Undaunted, Hale persuaded John D. Hooker, a wealthy Los Angeles businessman, to donate money to purchase the mirror blank. Andrew Carnegie ultimately provided the funds to build the rest of the project.

The 100-inch Hooker Telescope on Mount Wilson. Astronomer Edwin Hubble used this telescope to demonstrate that the Universe is expanding.
Ken Spencer/Wikipedia

After difficulties too numerous to describe here, Hale was finally able to look through the completed 100-inch telescope in 1917. The telescope was first pointed at Jupiter, and to his horror, Hale saw not a single sharp image but rather six or seven overlapping, blurry images. Was the telescope a failure? Then someone pointed out that workmen had left the dome open in the daytime, and the heat of the Sun might have distorted the mirror. Hale waited several hours in order to let the mirror reach the nighttime temperature of the surrounding air, and then looked at Vega. This brilliant blue star gleamed like a diamond—a sharp, clear image. The 100-inch was a success and would completely change our view of the Universe.

The triumph of the 100-inch telescope only made astronomers yearn for more. In 1921, Hale asked the designer of the 100-inch Hooker Telescope to prepare drawings of a potential 300-inch telescope. There were two insurmountable barriers to building such a large telescope. The first, as always at the start of a new telescope project, was lack of funds. Even more critical was lack of technology—after all, a 300-inch telescope wasn't built until the 1990s. As just one example, there weren't enough furnaces available to melt and pour so much glass into a mold, and even if there had been, the type of glass used for the 100-inch would require a human lifetime to cool slowly enough so that a 300-inch mirror would not crack.

Hale was not one to be daunted by practical difficulties. Corning Glass Works was experimenting with a type of glass—Pyrex borosilicate—that did not expand or contract very much when its temperature changed

Assembly of the yoke that holds the 200-inch Hale Telescope. The yoke was manufactured by Westinghouse. This 1938 image shows just how massive the 200-inch telescope is. Palomar/Caltech/Froebel Collection

rapidly by large amounts. That is why Pyrex is suitable for cooking, and Hale thought it could be used for large telescope mirrors. By 1928, Hale had again proved his powers of persuasion by obtaining construction funds from the Rockefeller Foundation for, once again, the world's largest telescope—although he had decided that building a 200-inch telescope was more prudent. Hale unfortunately did not live to see his vision realized. He died in 1938, a little more than a decade before the magnificent 200-inch Hale Telescope on Palomar Mountain in southern California was turned over to astronomers in 1948.

Mount Wilson Observatory Today

Mount Wilson is severely compromised by light pollution from the Los Angeles basin. Nevertheless, because of its very good image quality and easy accessibility, Mount Wilson remains a suitable place for observations of the Sun and bright stars. The newest facility on Mount Wilson is the CHARA interferometer. CHARA, which is operated by Georgia State University and several partner institutions, is an array of six telescopes, each of which is one meter in diameter. The telescopes are spread around the mountain in such a way that when the signals

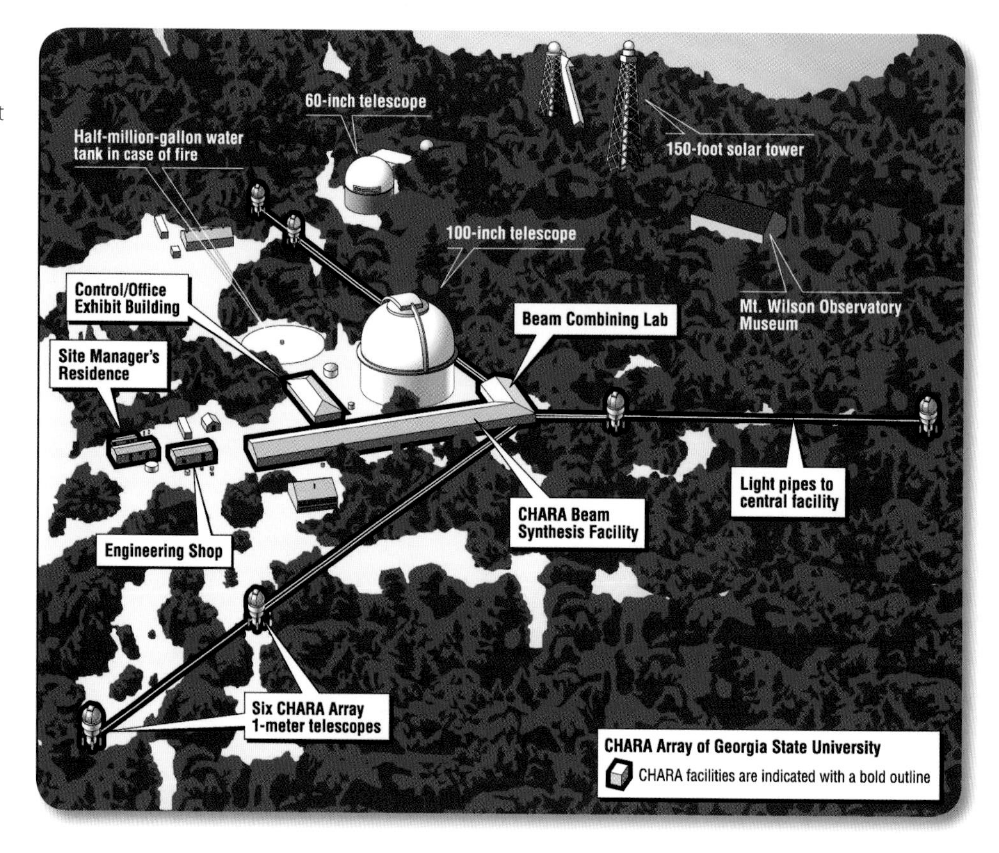

The top of Mount Wilson as it appears today. The many different telescopes are labeled. Note the three arms of the six-telescope CHARA interferometer.

Georgia State University

from all of them are combined, it is possible to see fine details that would otherwise require a single telescope with a diameter of a fifth of a mile. The ability to see fine details is called *resolving power*, and CHARA's resolving power is sufficient to resolve a nickel at a distance of 10,000 miles. This capability is useful for many types of observations, including, for example, measurements of the diameters of stars and the orbits of close binary stars.

Mount Wilson was seriously threatened by fire in 2009. In 2003, five of the six telescopes at Mount Stromlo in Australia were destroyed in a firestorm that raced through the surrounding pine forest.

For information on visiting Mount Wilson Observatory, go to www.mtwilson.edu.

Palomar Today

Unmatched in size and sensitivity for four decades, the 200-inch Hale Telescope has made a great many scientific discoveries. Especially notable are:

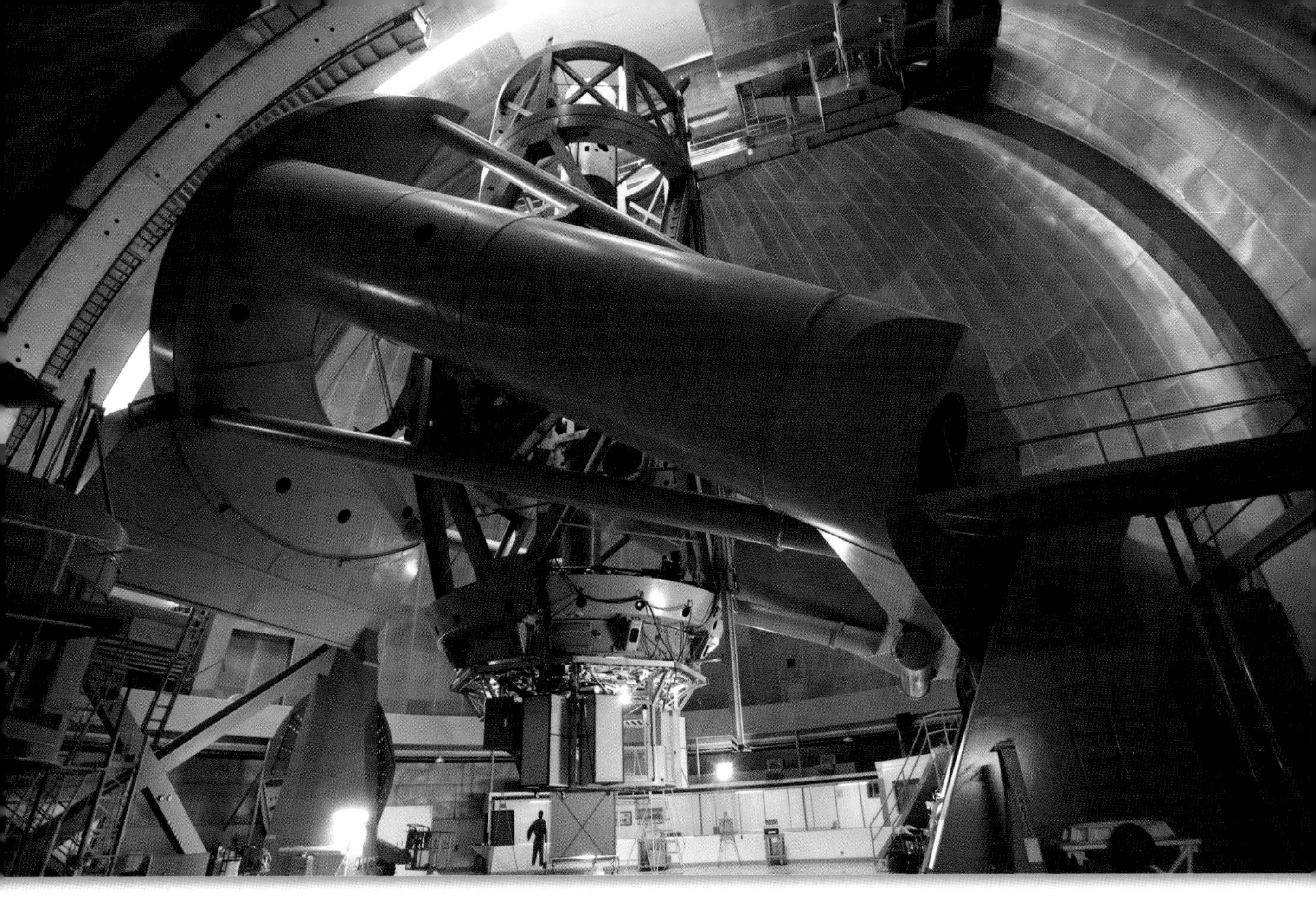

The 200-inch Hale Telescope. For scale, note the electrically heated flying suit on a stand near the bottom center of the image. Many years ago, astronomers working in the dome wore this suit to keep warm during long, cold winter nights. Now that sensitive digital cameras are available, astronomers can monitor their observations from a warm control room.
Caltech/Palomar Observatory

- Work on establishing the distances to remote galaxies and the rate of expansion of the Universe;
- Ground-breaking research on the evolution and composition of stars, which led to work that showed that all of the elements in the Universe except hydrogen, helium, and lithium are formed deep in the interiors of stars, then driven out into space through giant explosions (supernovae), only to be recycled into new generations of stars;
- The discovery of quasars, which turned out to be accreting black holes in the nuclei of galaxies; we now know that most galaxies harbor black holes with masses 1 million to 10 billion times the mass of the Sun in their centers (see Chapter 6).

In addition to the 200-inch Hale Telescope, two smaller telescopes are in operation at Palomar. The 48-inch robotic Samuel Oschin Telescope is currently used to scan large areas of the sky to search for transient events—objects that suddenly change in brightness such as supernovae. Dwarf planets in our solar system were discovered with

EXPLORING ON YOUR OWN

Observing the Sun, Moon, and Planets

Imagine yourself in the Southwestern desert 1,000 years ago. There are no artificial lights or clocks or calendars. The only way to communicate directly with neighboring villages is by walking from one to another to convey messages in person. How do you know when to hold great festivals or come together with your neighbors to trade or when to prepare your irrigation ditches to channel the summer rains to your crops?

The Ancestral Puebloans in the Four Corners area and their cultural kin, the Hohokam in Arizona, were skilled observers of the sky. Some of the walls of their buildings, notably Casa Bonita and Casa Rinconada in Chaco Canyon in northwestern New Mexico, were oriented with an accuracy of better than one degree to the true north-south line. Such accuracy could be achieved only through astronomical observations.

Perhaps the most remarkable discovery at Chaco is the so-called "Sun Dagger." High on Fajada Butte, there are three stone slabs standing upright. These stones are 6 to 10 feet high, 2 to 3 feet wide, 8 to 20 inches thick, and weigh about 4.5 tons. The slabs are separated by about 4 inches and stand perpendicular to the cliff immediately behind them.

At the summer solstice—when the Sun reaches its highest point at noon in the summer (on modern calendars usually on June 21)—sunlight makes its way through the 4-inch opening between two of the slabs and appears as a dagger shape on the back wall. A spiral petroglyph carved on the back wall is bisected by the dagger of sunlight for only about 14 minutes and only near the summer solstice. (Since this discovery in 1977, some of the stones have slipped so that the spiral is no longer bisected at the solstice. Because of the fragility of the site, it is no longer open to the public.)

At several other sites, including Hovenweep and Casa Grande, there are small openings in building walls that pass sunlight only at the summer or winter solstices or at the equinoxes. The winter solstice is the time when the Sun at noon is lowest in the sky (December 21) and the equinoxes in modern terms mark the beginning of spring (March 21) and fall (September 21). At the equinoxes, the Sun rises due east and sets due west. With these critical times established, one can imagine that the

▲ **The back wall of Chetro Ketl, one of the great houses in Chaco Canyon in northwestern New Mexico.** This very long, straight wall demonstrates the remarkable skill of Ancestral Puebloan builders. This wall is not aligned with the cardinal points of the compass, but in another great house, Pueblo Bonito, there is a long wall aligned within a degree of the true north-south line.

▲ **Casa Grande, located between Phoenix and Tucson, was built around 1350 by the Hohokam.** The original building was four stories high and had eleven rooms.

▲ **The back wall of Casa Grande.** The setting Sun can be seen from inside through the small, circular opening on the upper left wall at the time of the summer solstice. All of the walls were made of caliche (hardpan), which is a mixture of clay, sand, and calcium carbonate. The Hohokam softened the caliche with water so that it was about the consistency of a mud pie, and then kneaded it to a consistency of dough. It was then piled into a layer about 2 feet high and allowed to dry before the next layer was added. © Sidney Wolff

Ancestral Puebloans and Hohokam could schedule festivals or gatherings to trade goods for, as an example, the first full moon after the winter solstice. In this way, people from a sizable area could gather together. Chaco itself is now thought by some to be a place of pilgrimage.

In modern times, we have lost touch with the sky, but the kinds of observations made by these ancient people can be easily replicated. Fix a pole upright in the ground; when the shadow is shortest (in modern terms at noon), then it is pointing due north. Measure the length of the shadow at noon during the summer. When the noon shadow is shortest, that day is the summer solstice. Watch where the Sun rises or sets relative to fixed points on the horizon, such as a mesa or butte or, in modern cities, a building or tree. When the Sun rises farthest north, that is the summer solstice; the day it is farthest south is the winter solstice. Watch the Moon as it changes from a thin crescent to fully illuminated and back again during its 29.5-day cycle. Notice that at some times during the month you can see the Moon during the daytime.

Though we do not know how important stargazing was for the Ancestral Puebloan and Hohokam cultures, they must have observed Venus, which is the brightest object in the sky after the Sun and the Moon. Venus appears as an evening star just after sunset and at other times as a morning star just before sunrise. Mars, Jupiter, and Saturn are also as bright as the brightest stars. Look in your local paper or at websites maintained by *Sky and Telescope* and *Astronomy* magazines or on McDonald Observatory's StarDate website for maps that show which planets are above the horizon on the night you plan to observe. There are free apps for smartphones and tablets that indicate which planets can be observed from any specific location on any given date and time of night. These apps make it very easy to find the planets.

And do visit some of the Ancestral Puebloan sites—Mesa Verde, Chaco Canyon, Hovenweep, and several others are located in the Four Corners area. Casa Grande has an excellent visitor center that describes the Hohokam culture. ●

▲ **A passageway at Aztec Ruins National Monument in New Mexico.** Although not taken at a particular time astronomically speaking, it does show how impressive a beam of sunlight passing through a small opening would be in the darkened interiors of buildings like Casa Grande. Constructed between about A.D. 1085 and 1120, this site was vacated by about 1275, as was much of the Four Corners area. © Sidney Wolff

this telescope (see Chapter 4). The 60-inch telescope is used to make detailed follow-up observations of the discoveries made with the Oschin Telescope.

For information on visiting Palomar Observatory, go to http://www.astro.caltech.edu/palomar/visitor.

McDonald Observatory

Fort Davis, Texas

William J. McDonald was born in the Texas Republic in 1844, one year before Texas became part of the United States. He fought briefly with the Confederacy, and tried a number of professions after the war, including printing and the law. With money earned from legal fees, he was able to buy up undervalued assets during the depression of the 1870s. In the recovery that followed, McDonald became a rich man and a banker. He was introverted and unmarried but had many interests. His library included not only the English classics but also many books on botany, geology, and the geography and history of Texas. He also owned a small telescope, and perhaps his interest in the sciences explains his remarkable legacy.

McDonald died in 1926, and his will left nearly all of his estate to "the Regents of the University of Texas, in trust, to be used . . . for the purpose of aiding and equipping an Astronomical Observatory." The bequest to the university was more than $1,000,000, or about $13,000,000 in today's dollars. Naturally, his relatives were dismayed and in several lawsuits claimed that McDonald was of unsound mind and that his will should be declared invalid. Ultimately, the University prevailed but then faced the challenge of what to do with the money.

The University learned of the bequest initially from a reporter, and the dean at that time, Harry Benedict, said it was "like lightning from a clear sky." The University of Texas at that time had no working astronomers who could build or staff an observatory. Fortunately, in preparing for the court challenge to the will, Benedict had corresponded with Edwin Frost, the Director of Yerkes Observatory. Yerkes had the opposite problem—a talented staff of astronomers but no money to build a telescope larger than their 40-inch refractor. The Yerkes staff also wanted an observatory in a site better than southern Wisconsin and had already looked at possible sites in west Texas.

After extensive and difficult negotiations, the University of Texas and the University of Chicago, which operated Yerkes, reached an agreement to collaborate on the building of an observatory. At the time, such an agreement was unprecedented. Chicago was a prestigious university known for its research, while Texas was still a relatively obscure southern university—not the powerful institution it subsequently became. Partly because of this disparity, the agreement called for the first Director of McDonald Observatory to be Otto Struve, who was also Director of Yerkes. After rejecting sites near Austin, home of the University of Texas, and testing several mountains in Texas, astronomers selected Mount Locke, a 7,000-foot peak in the Davis Mountains in west Texas.

Struve made two important decisions about the telescope—ones that were controversial at the time but proved to be absolutely correct. First, the mirror blank would be made of Pyrex, the new type of glass chosen for the not-yet-cast 200-inch mirror. Pyrex has the advantage that it does not change its shape much as temperatures change. Temperatures do change during a night of observing, and a mirror that preserves its shape can continue to produce sharp images all night long.

Struve's second decision was to coat the mirror surface with aluminum. All glass mirrors must be coated with a thin layer of some kind of metal so that they reflect light efficiently. Traditionally, astronomers had used silver for their reflecting telescopes, but silver tarnishes quickly. Aluminum is much more durable than silver, and a new

The 82-inch mirror manufactured for McDonald Observatory. J. S. Plaskett, C. A. R. Lundin, and George A. Decker are viewing the mirror at Warner & Swasey Company in Cleveland prior to shipment to Texas. The telescope was later renamed the Otto Struve Telescope. This photo was likely taken in 1935. The mirror is small by modern standards, but at the time it was the second largest telescope mirror in the world. Warner & Swasey Company

A star party and the Milky Way under the dark skies at McDonald Observatory. The streak at top center shows the path of an artificial satellite. The Frank N. Bash Visitors Center has many exhibits and offers telescope tours and star parties. Ethan Tweedie Photography

technique had been developed for depositing aluminum on mirrors by placing them in a vacuum chamber. Struve's decisions proved to be good ones. All large telescopes are now made of low-thermal-expansion glass, and aluminum is the coating of choice for most mirrors.

In 1933, the Regents of the University of Texas signed a contract with Corning for an 81-inch mirror, the largest they could afford but also one that at that time would be the second largest in the world. Two attempts were required to make a successful blank. When the molten glass had cooled and solidified, it was discovered that the pressure of the glass had stretched the mold, and Texas now had an 82-inch mirror. It required several years to polish the mirror to the correct shape, and it finally arrived on Mount Locke in 1939. Photographs of the Pleiades and the Orion nebula demonstrated that, in Struve's words, the images were "perfect."

McDonald Observatory Today

Two larger telescopes have now joined the 82-inch Struve Telescope at McDonald. The 107-inch Harlan J. Smith Telescope, named after the Director of McDonald from 1963 to 1989, was funded by NASA to carry out studies of planets in our solar system as a complement to space missions. The 9.2-meter effective aperture (362-inch) Hobby-Eberly Telescope (HET), dedicated in 1997, is one of the world's largest (see Chapter 3). A new instrument for this telescope, HETDEX, will be used to map the Universe by collecting data on at least one million galaxies at distances of 9 to 11 billion light-years. HETDEX will observe fewer galaxies than DESI at Kitt Peak, but because the HET is a larger telescope, it can measure fainter and more distant galaxies.

For information on visiting McDonald Observatory, go to https://mcdonaldobservatory.org/visitors.

Summary

Each of these observatories was a pioneer in ways that influenced the development of subsequent observatories. Lick was the first to be built on a mountaintop in order to obtain clearer, sharper images than were possible at sea-level sites. In his choice of a site, Lowell put the quality of the observing conditions above more traditional considerations, such as easy access. Mount Wilson demonstrated that larger and more powerful telescopes could be built with reflecting mirrors rather than with lenses. McDonald

Observatory was the first to be built as a partnership of universities working together to achieve an ambitious goal. The 200-inch Hale Telescope on Palomar represented the state of the art for about four decades and also developed the technology that enabled the construction of several other telescopes, including the 4-meter Mayall Telescope at Kitt Peak.

Kitt Peak established the principle that *all* astronomers, independent of institutional affiliation, should have an opportunity to apply for observing time. This principle has been adopted by NASA, the European Southern Observatory, and other astronomical organizations. As a result, literally thousands of scientists have been able to contribute to the rapid advances in our understanding of the Universe during the past 50 years—a period of time that astronomers refer to as the "Golden Age of Astronomy."

The Hobby-Eberly Telescope. This telescope is the newest at McDonald and has an effective aperture of 9.2 meters. Thomas A. Sebring/McDonald Observatory

A **CLOSER** LOOK

Adaptive Optics

One of the advantages of observing from space is that astronomical images are crisp and clear. Here on Earth, the images of stars seem to flicker, and they appear blurry when seen through a telescope. The atmosphere is the culprit. When the light from a star moves through the atmosphere to our eyes and telescopes, it encounters atmospheric layers with different temperatures, different pressures, and different wind speeds. These differences create subtle changes in the path of the light that can distort stellar images and cause them to appear to dance around on time scales of a small fraction of a second. Astronomical exposures often take minutes to hours, and all of these small motions add up to a blurred image.

For decades, astronomers speculated that it might be possible to correct the blurring produced by the atmosphere and achieve sharp images on the ground. The technique for doing this is adaptive optics, and it is only in the twenty-first century that this technique has been widely used on telescopes. Adaptive optics is one of the new technologies being tested at Lick Observatory. Once proven at Lick, adaptive optics instruments are deployed to the much larger Keck telescopes in Hawaii.

▲ **A laser attached to the 120-inch telescope at Lick Observatory.** © Laurie Hatch

▲ **A laser beam shines high into the sky from the dome of the 120-inch telescope at Lick Observatory.** This beam produces an artificial star. Measurements of the artificial star by special instruments on the telescope can be used to calculate how much the atmosphere has distorted the image of the artificial star. This information can then be used to correct nearby images of real stars. You have probably seen the twinkling of stars near the horizon. Adaptive optics, in effect, puts a stop to the twinkling. © Laurie Hatch

There are two key components to an adaptive optics system. First, there must be a way to measure the distortion produced by the atmosphere. Because these distortions occur very rapidly, they must be measured hundreds of times each second. The distortions can be measured by observations of a bright star that is very near the area of the sky that an astronomer wants to observe. The star must be bright, because the measurements must be made so rapidly.

Unfortunately, not every patch of sky that an astronomer might want to observe has a bright star nearby. To be able to observe the entire sky, astronomers create an artificial star. One technique involves using a laser to excite sodium atoms high in the atmosphere. These sodium atoms then emit light in a small spot in the sky and serve as an artificial star. The light from this artificial star is then observed back at the telescope and analyzed, hundreds of times each second, to calculate exactly how the image has been distorted.

The second key component in an adaptive optics system is a deformable mirror. To understand how a deformable mirror works, think of a pliable piece of rubber stretched over a round hoop. Push on the rubber from underneath with your fingers. By moving your fingers up and down, you can push the rubber surface into a variety of complicated shapes. A deformable mirror also has a flexible surface with tens to thousands of actuators behind it. These actuators act just like your fingers, except that they can pull as well as push on the mirror. The observations of a star, either real or artificial, are used to calculate exactly what the actuators must do in order to produce a mirror shape that corrects for the distortions produced by the atmosphere.

Adaptive optics has been used to study crowded star fields, to image planets around stars, and to observe weather on the planets in our solar system. ●

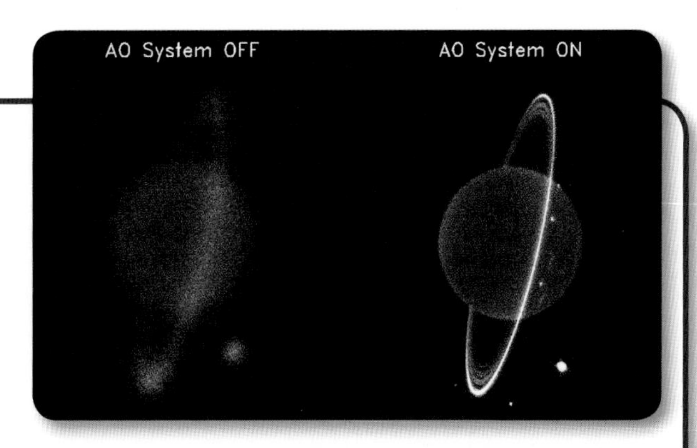

▲ **An image of Uranus taken with and without adaptive optics.** The improvement is remarkable. Heidi Hammel, Space Science Institute, Boulder, CO; Imke de Pater, University of California, Berkeley; W. M. Keck Observatory

▲ **Images of the center of the Milky Way Galaxy with (right) and without (left) adaptive optics.** Sharp images like this one over many years have been used to measure the orbits of stars very close to SgrA*, which is at the Galactic center. The stars are moving very fast (7,500 miles per second or about 4 percent of the speed of light). Such high speeds are possible only if the stars are orbiting an object with a mass 4 million times larger than the mass of the Sun crammed into a volume with a diameter not larger than the diameter of the orbit of Uranus. The only known astronomical object that could have so much mass in such a small space is a black hole. (See black holes on page 118). This image was created by Professor Andrea Ghez and her research team at UCLA and is from data sets obtained with the W. M. Keck Telescopes.

▼ **Before and after images of a binary star with an adaptive optics system (MagAO) on the Magellan Telescope.** This is the sharpest image ever taken at this wavelength, and it is actually sharper than can be achieved with the Hubble Space Telescope at this wavelength. The Magellan Telescope has a primary mirror with a diameter of 6.5 meters. The primary mirror of the Hubble Space Telescope is only 2.4 meters. Larger aperture means greater resolving power, i.e., better ability to see fine details in an image. Until the invention of adaptive optics, turbulence in the atmosphere blurred astronomical images (left image) and kept ground-based telescopes from achieving their full potential. Adaptive optics removes the blurring. Laird Close, University of Arizona

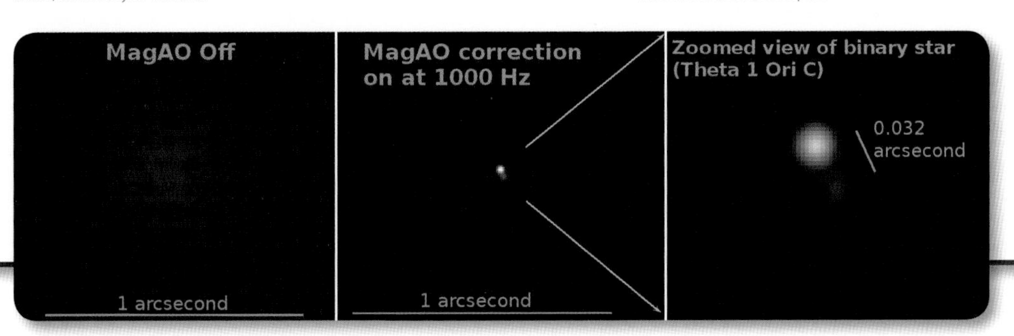

Gemini North 8-meter telescope and dome. This image was made by digitally stacking images of the open Gemini dome as it rotated. Approximately forty individual 5-second exposures were stacked to produce the transparent dome effect. One additional one-minute exposure was obtained of the sky after the dome was rotated to block the light in the interior. Gemini North is located on Mauna Kea in Hawaii. Gemini Observatory/AURA

2 Observatory Sites

Arizona when it was still a territory, a plateau at an altitude of 16,400 feet in the Andes, even the South Pole—why do astronomers build their state-of-the-art observatories in such remote, challenging locations? The observatories described in the previous chapter are the most scientifically productive optical observatories in the continental United States. Why is the Southwest the best region in North America for ground-based observations? And what has prompted astronomers to build new observatories with even larger telescopes in more remote places and at much higher altitude?

Observatories in the Southwest United States

Selection of sites for the three oldest observatories—Lick, Mount Wilson, and Lowell—was based typically on only a few nighttime observations that showed these sites were good enough for astronomy. Mountaintops were desirable because they were above local haze and fog. Lick had the additional attraction of being visible from property owned by James Lick. Douglass chose a site near Flagstaff for Lowell Observatory mainly because he had better luck with the weather than during his brief stops in southern Arizona. Mount Wilson was one of two mountains in southern California that Hale thought might be suitable and was chosen largely because it was easier to reach than Palomar Mountain, which was the alternative. In none of these cases was a serious attempt made to compare the relative merits of various potential peaks.

When the time came to select a site for the 200-inch telescope, criteria included:

- Latitude between 30 and 35 degrees in order to observe a large portion of the southern hemisphere sky.
- An altitude between 6,000 and 8,000 feet. The minimum altitude was set to guarantee that the atmosphere above the site was free from haze. A higher altitude was likely to have significant snowfall, thus making access difficult in winter.

- A large percentage of clear nights.
- A small range in temperature and low winds in order to minimize atmospheric turbulence and ensure good "seeing." When astronomical images are sharp, astronomers say that the "seeing" is good.

Sites in the southwestern United States and northern Mexico were the only ones close to Hale's office in Pasadena that could meet all these conditions. Hale had long been intrigued by Palomar but had rejected it in favor of Mount Wilson thirty years earlier because of Palomar's relative inaccessibility and distance from a large city that could provide the necessary logistical and technical support. The widespread use of the automobile and the growth of San Diego erased these negatives by the early 1930s, and Palomar was chosen for the 200-inch.

There were some additional criteria that influenced the choice of Kitt Peak. The national observatory had to be located in the continental United States so that astronomers from all over the country could reach it easily. To facilitate travel, an airport had to be nearby. The site had to be far enough away from city lights to avoid light pollution but within a reasonable distance of an urban environment so that the staff would have access to housing and schools and to a university where scientific, technical, and cultural exchanges could take place.

A further requirement was that the national observatory be located at a site with the maximum possible number of cloud-free nights to take advantage of the development of a new observing technique—photoelectric photometry. Until the early 1950s, quantitative measurements of astronomical sources relied on photography. To make an observation, an astronomer exposed a photographic plate, sometimes for a few minutes or, if the goal of the research program was to detect very faint sources, for several hours. He or she then developed the plate in a dark room.

While photographic plates were the best available technology for many decades, they had several disadvantages. Light can be described as consisting of bundles or packets of energy called photons. Photographic plates are actually not very sensitive to light and can detect only about 1 percent of the photons that reach them. Conversion of the information on the plates to, for example, quantitative measurements of the brightness of the stars was difficult, tedious, and had a typical accuracy of only 5 to 10 percent.

In the early 1950s, electronic devices called photomultiplier tubes (PMTs) became commercially available. These were much more sensitive to light than photographic plates and much easier to calibrate, resulting in measurements with an accuracy of 1 percent or better—a considerable improvement. Such accuracy can be achieved, however, only if the sky is completely free of clouds. Even clouds so thin that they cannot be seen by eye at night can make accurate measurements impossible. Furthermore, it is important that *all* of the sky be clear all night long. Otherwise it is impossible to compare the brightness of a star in one part of the sky with the brightness of a different star in another part of the sky or to study the changes in brightness throughout the night of a variable star.

PMTs have been replaced by charge-coupled devices (CCDs) very similar to those in digital cameras and smartphones—but optimized for detecting faint astronomical sources and making accurate measurements of brightness. Even for CCD observations, a high premium is placed on finding sites with a large percentage of clear nights.

To find the site with the largest number of clear nights for the national observatory, astronomers examined maps provided by the Department of Agriculture. The region of the United States with the largest number of clear days, according to these maps, was centered on Yuma, Arizona. Kitt Peak is located 260 miles east of Yuma and does not have quite as many clear days, but it is clear more often according to these maps than any of the other sites already developed in California and the Southwest. (The maps indicated only clear *days* since no one had observed the number of clear *nights*, but except during summer monsoons, the weather is usually similar.) Kitt Peak also met the other criteria established for the national observatory.

The search for a site for the national observatory was restricted to mountains with an elevation of less than 8,000 feet and in the United States to provide easy access for visiting astronomers. Yet in 1962 the decision was made to build a second national optical astronomy observatory in Chile. And in 1968, construction of a major new observatory was underway on Mauna Kea on the Big Island of Hawaii. The summit of Mauna Kea is at an altitude of 13,796 feet. The next two sections describe why astronomers were suddenly willing to go so far afield.

Chile

The southern skies contain many unique objects that are not visible from northern hemisphere observatories. These include the Magellanic Clouds, which are two nearby galaxies, and the brightest globular clusters, which contain the oldest stars in our own Milky Way Galaxy. The center of the Milky Way Galaxy itself is much higher and more easily observed in the southern sky. Beginning in the nineteenth century, northern observatories had sent several expeditions to the southern hemisphere, and especially to Chile and South Africa.

It was, however, the Boeing 707 jet, which began providing service in 1958, that finally made it possible to consider establishing a major observatory in South America to

host visiting astronomers. In 1962, the United States decided to build Cerro Tololo Inter-American Observatory on a site about 300 miles north of Santiago, Chile. In 1963, the European Southern Observatory (ESO) opted to build its international observatory on La Silla, which is north of Cerro Tololo on the southern edge of the Atacama Desert.

In 1969, the Carnegie Institution for Science established the Las Campanas Observatory near La Silla. Las Campanas supplanted Mount Wilson, which was strongly compromised by light pollution, as Carnegie's main observing facility.

Chile is now home to several more major observatories because it offers some of the best sites in the world. The average number of photometric nights at La Silla is 63 percent. ESO's VLT, which consists of four 8.2-meter Very Large Telescopes, is located even deeper in the Atacama Desert and sees clear nights 74 percent of the time. Even the best sites do have clouds sometimes.

The sites in Chile, like Mauna Kea in Hawaii (see next section), have excellent seeing. *Seeing* is a measure of the quality of images as seen through a telescope. Bad seeing causes astronomical images to blur, move, or vary in brightness. You have probably seen the effects of bad seeing. For example, if you look over the top of a bonfire, objects beyond the bonfire will appear to shimmer and shake. This effect is the result of atmospheric turbulence, which is caused when parcels of warm air encounter cooler air.

You may have seen the twinkling of a star near the horizon. This, too, is the result of atmospheric turbulence. The seeing at an observatory is good only if atmospheric turbulence is low. The main factor in determining whether an observatory has good seeing is the topography upwind of the site. Images are sharpest at locations where the prevailing winds arrive after having crossed many miles of ocean. Because the ocean has a fairly uniform temperature, the airflow above it tends to be laminar with little turbulence. For this reason, the very best observing sites are on isolated mountains located in the middle of an ocean (e.g., Mauna Kea in Hawaii) or in coastal ranges (e.g., the observing sites in Chile).

The Milky Way as seen from Cerro Tololo Inter-American Observatory in the foothills of the Andes in Chile. The 4-meter Blanco Telescope is at the lower right. The Large Magellanic Cloud, a nearby galaxy, can be seen just clearing the horizon at the lower left. The dark rectangular feature in the middle of the Milky Way is known as the Coal Sack. The two bright stars to the left and up from the Coal Sack and just below the next dark cloud are Alpha and Beta Centauri. An imaginary line drawn between these two stars points down and to the right to the Southern Cross, which is just to the right of the Coal Sack. K. Don and NOAO/AURA/NSF

The observatories on Mauna Kea, Hawaii. In the foreground, from left to right, are the University of Hawaii (UH) 0.6-meter telescope at the far left, the United Kingdom Infrared Telescope (UKIRT), the UH 2.2-meter telescope, the Gemini North 8-meter telescope (silver, open), and the Canada-France-Hawaii Telescope. On the right are the NASA Infrared Telescope Facility (silver) and the twin domes of the W. M. Keck Observatory; behind and to the left of them is the Subaru Telescope. In the valley below are the Caltech Submillimeter Observatory (silver), the James Clerk Maxwell Telescope (white, open), and the assembly building for the submillimeter array. UKIRT is currently being funded by NASA and operated under a Scientific Cooperation Agreement among Lockheed Martin Advanced Technology Center, the University of Hawaii, and the University of Arizona.

The cinder cone with no telescopes on it in the center of the photograph is Pu'u Poli'ahu. In the distance is the dormant volcano Hualalai (altitude 8,271 feet). Richard Wainscoat/Gemini Observatory/AURA/NSF

Mauna Kea, Hawaii

In the early 1960s, NASA decided to build a ground-based telescope to obtain observations in support of their space program of planetary exploration. Gerard Kuiper of the University of Arizona was commissioned to conduct a site survey in Hawaii. In early 1964, he and his staff set up a site-testing dome on Mauna Kea to measure the seeing, which proved to be excellent.

Although Mauna Kea was exceptionally good for astronomy, there was substantial risk involved in building and operating an observatory at an altitude of nearly 14,000 feet. No one knew how astronomers would perform at such a high site. After all, pilots of unpressurized aircraft must use oxygen continuously any time they are above 10,000 feet for more than 30 minutes and all the time when they are above 12,000 feet. Haleakala on Maui at 10,000 feet elevation offered a far more convenient site because it was already developed with roads and a power line. In contrast, Mauna Kea had no source of power, access was by a very primitive dirt road negotiable only by four-wheel-drive vehicles, and winters were accompanied by high winds and heavy snowfall.

Comparison site testing of Mauna Kea and Haleakala showed, however, that in strictly astronomical terms Mauna Kea was superior in every way. John Jefferies of the University of Hawaii made the bold decision to build the 88-inch telescope funded by NASA on Mauna Kea. Once the feasibility of working at that altitude was demonstrated, many other countries followed suit, and Mauna Kea is now widely recognized as the best astronomical observing site in the northern hemisphere.

Infrared Astronomy

Why the sudden interest in high-altitude sites? The observatories described in Chapter 1 were built as optical observatories. But astronomical sources emit energy that our eyes are not sensitive to—from gamma rays and X-rays to radio waves. Scientists call all of these forms of energy "electromagnetic radiation." Electromagnetic radiation has nothing to do with radioactivity; radiation in this context refers to the emission of energy in the form of waves.

Astronomers needed to go to high altitude to explore the new field of infrared astronomy. Infrared radiation is a form of electromagnetic radiation that has longer wavelengths than visible light. Humans sense infrared radiation as heat. Stick your hand in a hot oven, and you will sense infrared radiation. Longer wavelengths, however, mean lower energy, and the energy associated with infrared radiation was so low that it did not affect the photographic plates or even the photoelectric photometers used by astronomers through the 1950s. Infrared astronomy became possible only in 1961 when Frank Low, who spent most of his career at the University of Arizona, developed a device called a bolometer that changed its electrical resistance when it absorbed infrared energy. Essentially a thermometer, Low's device proved to be sensitive enough to measure the infrared heat energy emitted by stars and other astronomical sources.

After several decades of development, we now have infrared detectors that, like CCDs, have thousands of pixels. The

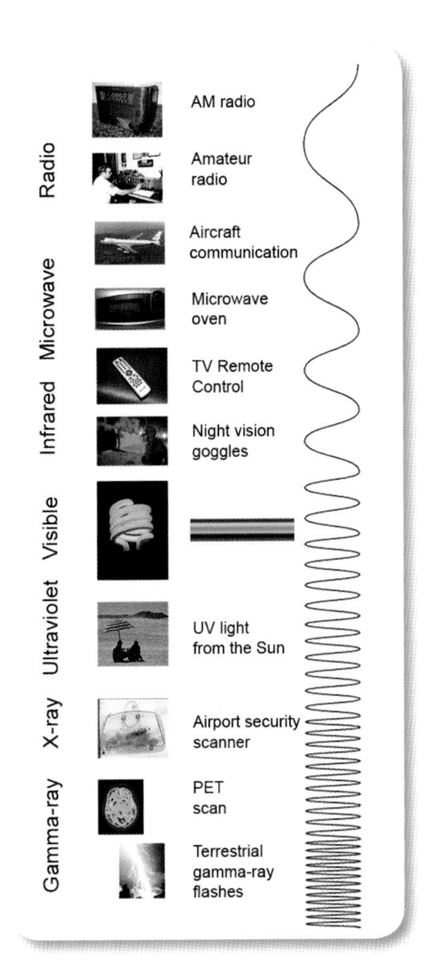

The electromagnetic spectrum from the lowest energy/longest wavelength (at the top) to the highest energy/shortest wavelength (at the bottom). The small images show applications on Earth of the energy associated with each wavelength. Astronomical sources, including gases, stars, and galaxies, emit energy at all wavelengths of the electromagnetic spectrum. Much of the progress of modern astronomy is due to the fact that we now have facilities, either on the ground or in space, that can detect all of these wavelengths.

NASA's Imagine the Universe

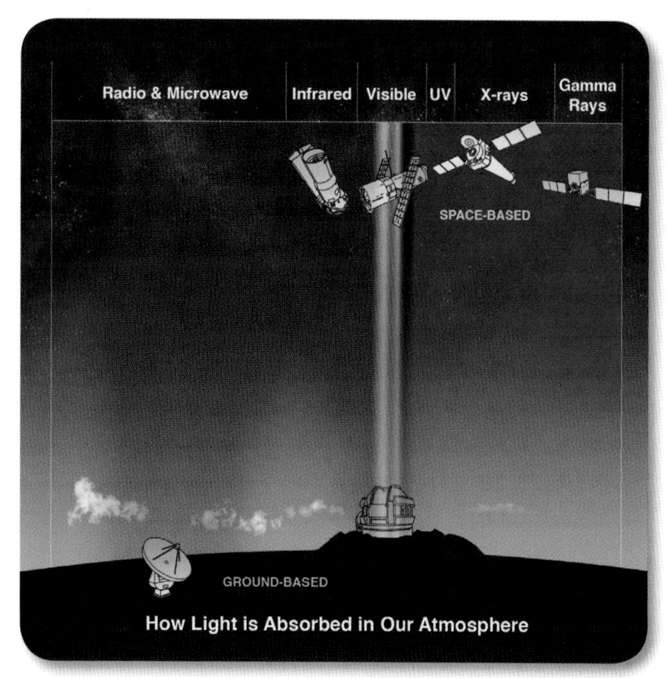

The Earth's atmosphere absorbs most types of electromagnetic radiation from space before it reaches the Earth's surface. This illustration shows how far into the atmosphere different parts of the electromagnetic spectrum can penetrate before being absorbed. Only portions of the radio spectrum and visible light reach the surface.

NASA/CXC/M. Weiss

images made with infrared cameras have been key to our understanding of the processes of star formation and the evolution of galaxies.

With suitable infrared detectors in hand, the next challenge faced by astronomers was the Earth's atmosphere. Fortunately for us, the atmosphere absorbs much of the short-wavelength electromagnetic radiation that reaches the Earth from space. For example, the ozone layer in the lower stratosphere (12–19 miles above the Earth's surface) absorbs much, but not all, of the ultraviolet electromagnetic radiation from the Sun. If there were no ozone layer and more ultraviolet reached the Earth's surface, we would be at much greater risk for sunburns. This type of radiation can also cause genetic damage that results in diseases such as skin cancer.

Unfortunately for astronomers, the atmosphere also absorbs infrared radiation. In this case, however, the absorption is due to water vapor, which is concentrated at low altitudes. Site survey measurements show, for example, that the amount of water vapor above Mauna Kea at 13,800 feet is on average only about half that of a mountain in Chile at 7,500 feet. It was primarily in order to exploit the potential of the new field of infrared astronomy that observatories on Mauna Kea were built.

In their search for ever-drier locations, astronomers are now working at sites even more inhospitable than Mauna Kea. Several astronomical experiments have been carried out at the South Pole. In the winter, when it is dark for six months, the average temperature is –72 degrees Fahrenheit. Odd as it may seem, the South Pole has a desert climate, rarely receives any precipitation, and the humidity of the air is near zero.

Slightly more welcoming to humans is the site of the newly completed Atacama Large Millimeter Array (ALMA), a complex of 66 antennas designed to detect millimeter and submillimeter radio waves. Atmospheric water vapor absorbs millimeter radiation even more efficiently than it absorbs infrared radiation, and so ALMA has been built in the Andes of Northern Chile on a plateau at an altitude of 5,000 meters (16,400 feet). This site is extremely dry, cloud cover is nearly nonexistent, and there are no nearby cities to cause any interference with the detection of radio signals from space.

It is very expensive to operate at remote sites, at high altitude, or at the South Pole. Therefore, any observations that do not require such extreme conditions should be made at sites where it is easier and less expensive to work. As we have seen, all of the Southwest

The Atacama Large Millimeter/submillimeter Array (ALMA), located at 5,000 meters (16,400 feet) on the remote Chajnantor Plateau in the Chilean Andes. Some of the 66 antennas are visible here. The spear of light was produced by a meteor, which is a small bit of rock from interplanetary space, as it burned up in the Earth's atmosphere. The two bright objects just above the antennas in the center of the picture are Spica, a star in the constellation Virgo, and Mars.
ESO/C. Malin

observatories continue to make important discoveries. Lick has focused on the search for planets and supernovae; Mount Wilson has a state-of-the-art optical/infrared interferometer; the Lowell and McDonald Observatories have recently completed new telescopes; and Kitt Peak's Mayall and WIYN Telescopes will soon be initiating surveys designed to analyze the properties of dark energy and exoplanets.

Light Pollution

All of the observatories in Arizona and California have been negatively affected at least to some degree by light pollution from nearby cities. The International Dark-Sky Association (IDA) defines light pollution as any adverse effect of artificial light, including sky glow, glare, light trespass, light clutter, decreased visibility at night, and energy waste. Astronomers are particularly concerned with sky glow, which is produced by streetlights and other lighting fixtures that emit a portion of their light upward, where it is scattered by the atmosphere and produces an orange-yellow glow above a city or town. This glow makes it difficult or even impossible to detect faint astronomical sources.

A **CLOSER** LOOK

Infrared Astronomy

Infrared energy is sensed as heat energy. If we could look at the world with infrared eyes, it would look very different. These images show flamingos as our eye sees them (top) and as they appear in an image taken by an infrared camera (bottom). Feathers act as good insulators, and so the bodies of the flamingos appear relatively cool. The hottest spots are around the eyes. Also, note that some have one leg that is far hotter than the other. Flamingos lose body heat very quickly through their legs, so they often stand on one leg, with the other tucked up next to their body, keeping it warm. The flamingos we see with one hot and one cold leg have been standing on the cold one for a while, and have just gone back to standing on both.

Just as you can sense the heat of the Sun on a very smoggy day even though the Sun itself may look dimmer to the eye, so too can infrared light penetrate

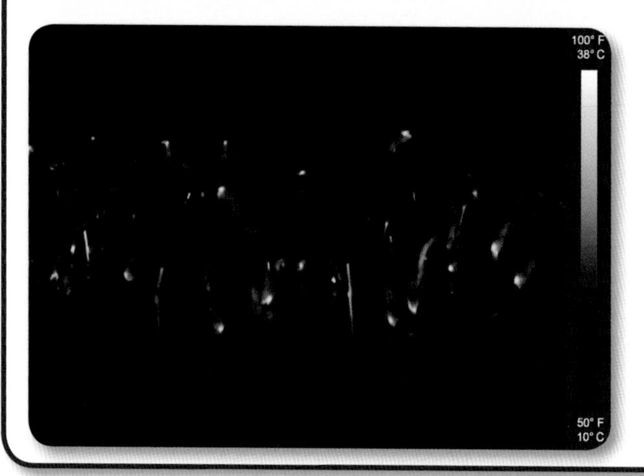

▲ **Visual image.**

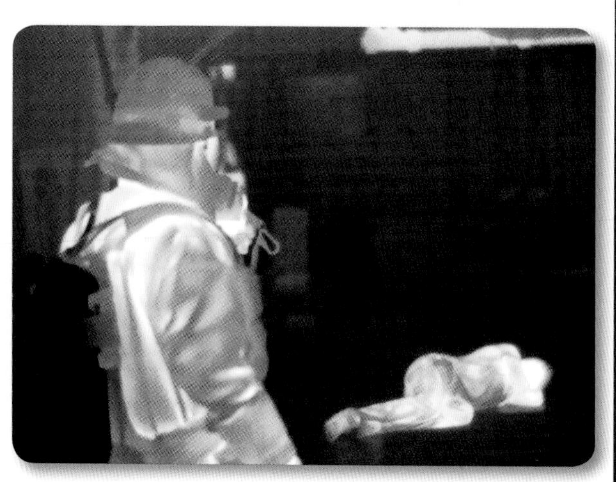

▲ **Infrared image.** Cool Cosmos/IPAC (all photos on this page)

through the dense smoke of fires. Firefighters routinely use portable infrared cameras to navigate through smoky areas.

Stars form in clouds of dust and gas. When astronomers could observe only at visible wavelengths of light, much of the action was hidden from view. With infrared cameras, astronomers can penetrate dust clouds and observe newborn stars. ●

◀ **The Trifid Nebula seen in visible light (top) and infrared light (bottom).** The Trifid is a giant star-forming cloud of dust and gas. Dark lanes of dust are seen in the visible image. The blue light in the visible image is the scattered light from a hot blue star. The red light is emission from hydrogen atoms. Dust obscures the optical light emitted by stars embedded within the dust cloud, but some of these young stars can be seen in the infrared. The warm dust clouds also emit heat energy and glow in the infrared. The infrared image is what astronomers call a false-color image. Blue colors show the hottest regions. Red are the coolest, and green and yellow regions have intermediate temperatures. Top: NOAO/AURA/NSF. Bottom: image and comparison created by Spitzer Space Telescope/IPAC/NASA

Light pollution is not an issue just for astronomers. Humans have enjoyed the beauty of the dark night sky for thousands of years, but that beauty has been lost for those who live in large cities. Excessive lighting can also affect living systems, such as human sleep patterns and bird migration. The challenge is to design outdoor lighting that achieves the illumination required for safety while not using more light than is necessary.

The website of the IDA (www.darksky.org) is a rich source of information about exactly how to achieve good outdoor lighting. It includes slide presentations, lesson plans for students, and a model lighting ordinance.

Some types of lighting are friendlier to astronomy than others. Low-pressure sodium lights, which put out a nearly pure yellow color, are especially favored near observatories because these lights affect only a small part of the optical spectrum. The other colors of light from blue to red are not affected. White lights put out all of the colors from blue to

The United States at night. NASA Earth Observatory/NOAA NGDC

red and are particularly undesirable. Shielding light fixtures so that the light is directed downward where it is needed rather than up into the sky where it can produce sky glow is also important. The IDA has worked closely with the Illuminating Engineering Society to develop a model lighting code that identifies types of lighting products that are energy efficient and fully shielded.

The ongoing revolution in lighting is presenting a new challenge to the task of preserving dark skies. Cities are beginning to change their street lighting to LEDs (light-emitting diodes). Although LEDs cost more than conventional lighting, they last longer, are more energy-efficient than incandescent and sodium lighting, and do not use materials like mercury that are harmful to the environment. The downside of many of the LEDs currently being installed is that they emit more blue light than do sodium lights. Because the atmosphere scatters blue light more efficiently than yellow and red light, LEDs will increase sky glow, which is a problem for astronomers. The increased blue light at night may also affect human sleep patterns, bird migration, and other living systems. In order to earn the IDA's Fixture Seal of Approval, light fixtures must limit the amount of blue-rich light emitted into the nighttime environment by using filters.

Good lighting ordinances do work! Phil Massey, a Lowell Observatory astronomer, made observations of the sky brightness over Kitt Peak in 2009 and 2010 and compared the results with measurements he made 10 years earlier. The results showed very little change in the sky brightness even though the population of Phoenix, Tucson, and the surrounding counties increased by about 23 percent. Light pollution has increased at a rate much less than population growth—thanks to effective lighting codes.

EXPLORING ON YOUR OWN

How Bright is Your Sky?

The "Globe at Night" is an ongoing citizen science project to measure sky brightness around the world. "Citizen science" is scientific research conducted by amateur or nonprofessional scientists. It is easy to participate, and this is also a good way to get to know some of the most important constellations. There are free apps for smartphones and tablets that will help you identify the constellations.

The website (www.globeatnight.org) describes the five easy steps required to estimate the sky brightness wherever you are:

1. For each month of the year, the Globe at Night website gives a map of a constellation that is easily visible and specifies the brightness of the stars in that constellation.
2. If you want to report your results, the Globe at Night website can also be used to find the latitude and longitude of the location where you are making your observations.
3. Go outside more than an hour after sunset (8–10 p.m. local time). The Moon should not be up. Let your eyes get accustomed to the dark for 10 minutes before your first observation.
4. Match your observation to one of seven magnitude charts for the constellation that is observable that month and note the amount of cloud cover.
5. Use the website to report the date, time, location (latitude/longitude), the star chart you chose, and the amount of cloud cover at the time of observation. You can then compare your results to thousands around the world! If you make observations from different locations—city center, suburbs, a dark rural location—you can see the effects of artificial lights on sky brightness. ●

Dark Skies over the National Parks

If you have done the Globe at Night experiment from your backyard, you may be disappointed in the small number of stars you can see. Most of us live in urban areas where artificial lights dominate the night sky. Estimates are that only about 10 percent of Americans live in places dark enough to see the estimated 2,500 stars that were visible at, for example, Chaco Canyon when the Ancestral Puebloans lived there.

The Southwest and the Colorado Plateau offer ideal conditions for observing the stars. As the image of the United States at night shows, the Southwest is one of the most sparsely populated areas of the United States. Brightly lit large cities are few and far between. Air pollution is low, and most nights are free of clouds. As an added bonus, the Southwest is blessed with many national parks, which limit artificial lighting and protect the dark skies that arch over them.

As one of its programs, the International Dark-Sky Association (IDA) certifies parks that meet specific standards required to earn designation as International Dark Sky Parks. To date, twenty parks in the United States have received certification. These are parks or public lands with "exceptional nighttime beauty, dark skies education, and preservation of the nighttime environment." This means that the sky is extremely dark, observers are not distracted by glare from artificial lights, and the full range of sky phenomena, including the Milky Way and faint meteors, can be observed. Fittingly, two of these sites are Chaco Culture National Historical Park in New Mexico and Hovenweep National Monument on the border of Utah and Colorado—both places where the Ancestral Puebloans are known to have observed the sky. A list of Dark Sky Parks in the United States is given at the end of this book. Check the IDA site for new additions to the list and for Dark Sky Parks in other countries (www.darksky.org).

Many parks, not just the ones listed in the back, offer ranger-led programs about astronomy. Grand Canyon, Bryce, Capitol Reef, Joshua Tree, and Great Basin have nighttime programs at least weekly in season. The typical nighttime program includes an interpretative presentation, identification of some of the more prominent constellations, and an opportunity to look at planets, star clusters, or other interesting objects through small telescopes. Many other parks also offer astronomy programs.

◀ **Thor's Hammer in Bryce Canyon National Park, Utah.**

© Royce Bair

Bandelier, Bryce, Death Valley, Great Basin, Great Sand Dunes, and White Sands schedule night hikes when the Moon is full. Some parks, including Grand Canyon, Bryce, Chaco, Parashant, Great Basin, and Capitol Reef, host annual astronomy festivals. National parks are also a great location for astrophotography. Search "images .google.com" for "national parks astrophotography," and you will find many spectacular photographs that combine an interesting rock formation in the foreground with a starry sky, often including the Milky Way, in the background. Arches National Park is a particular favorite with astrophotographers.

The next time you visit a national park, check the park website or flyers for astronomy programs. The most memorable part of your visit may begin after sunset! ●

◀ **Delicate Arch, Arches National Park, Utah.** © Phillip Colla

Star trails around Polaris, the North Star, above the Hobby-Eberly Telescope at McDonald Observatory in Texas. The HET has an effective aperture of 9.2 meters and is an example of a telescope that makes use of innovative technology to achieve very large light-gathering power.

Ethan Tweedie Photography

3 Ground-Based Telescopes—Bigger and Better

When Galileo first looked through a (very small) telescope in 1610, he made observations that clearly supported Copernicus's revolutionary ideas about the solar system—namely that the planets orbit the Sun, not the Earth. Galileo discovered the four brightest moons of Jupiter and observed their orbits, which proved that not all objects revolve around the Earth. He saw that Venus had phases, just as the Moon does, which could only be explained if Venus orbits the Sun rather than the Earth. Galileo also saw mountains and valleys on the Moon, which suggested that the Earth itself might be simply another planet—and not the center of the Universe.

Telescopes are the key to discovery in astronomy, and we live in a golden age of telescope building. New technologies are making it possible to build ground-based telescopes of a size that was unimaginable even 50 years ago.

What Telescopes Do

The main purpose of a telescope is to collect light from an astronomical source and then focus it into an image that can be observed with the eye or a camera. As an analogy, imagine that it is raining, and you set out a bucket and a garbage can to catch the water. Because the garbage can has a bigger diameter, it will collect more water. Just so, a larger telescope will collect more light than a smaller one and so can observe much fainter objects. The study of ever-fainter objects has been the key to understanding how the Universe has changed over time.

There are two main types of telescopes: refracting and reflecting. Refracting telescopes are, like modern binoculars and Galileo's telescope, made with lenses. Until the beginning of the twentieth century, refractors were the usual choice for astronomical research. Two large refractors in the Southwest are the 24-inch refractor at Lowell and the 36-inch refractor at Lick. As these examples show, telescopes are always described in terms of their largest optical element.

Cleaning the lens of the 36-inch Lick refractor. This lens is mounted at the top end of the telescope tube seen in the previous image. Because light must pass through the lens, the glass must have no imperfections. © Laurie Hatch

Three technical challenges prevented the manufacture of even larger lenses for refracting telescopes. First was the challenge of polishing the glass to the correct shape. The lens in your camera is made of more than one piece of glass. In a similar fashion, the Lick lens has two pieces of glass. Each side of each piece of glass—four surfaces—must be polished accurately to ensure that the resulting image is sharp. Second, since light must be able to pass through the lens, the glass must be perfect with no flaws or bubbles. Such perfection is hard to achieve in very large lenses. Third, a lens can only be supported from the edges, and the larger the lens, the more it sags due to gravity. Because of these challenges, the largest refracting telescope ever built is the 40-inch Yerkes Telescope in Wisconsin.

Scientists realized very early on that telescopes made with mirrors—reflecting telescopes—could avoid these problems. Only the front surface of the mirror needs to be polished to the correct shape. Mirrors can be supported from behind because light does not need to pass through them. For the same reason, flaws and bubbles in the interior of the glass can be acceptable.

The 36-inch refractor at Lick Observatory. The slot in the dome is open so that the telescope can view the sky. In a refractor, the light enters the top end of the telescope and, just as is true of binoculars, the light travels through the tube to an eyepiece or, in the case of a research telescope, to an instrument mounted at the bottom end. © Laurie Hatch

Isaac Newton is credited with making the first reflecting telescope in 1668. Its metal mirror was only about 2 inches in diameter. Technical problems, however, limited the popularity and scientific use of reflectors for the next two centuries. Speculum metal, which is a white, brittle alloy of copper and tin, was the only substance available at that time that could be ground and polished to the precise shape required to bring starlight to a focus. But speculum metal was hard to cast and shape accurately. When they were new, speculum mirrors reflected only about 66 percent of the light that reached them, and they tarnished quickly. This meant that speculum mirrors had to be removed, repolished, and reshaped frequently.

The era of reflecting research telescopes was initiated in the mid-nineteenth century with the development of a process for depositing a thin layer of silver on glass. Silver reflects about 90 percent of the light and does not tarnish as fast as speculum. The advantage of glass is that it holds its shape better than speculum as the temperature changes, which means that images remain sharp throughout the night. The first large reflector for astronomical research was the 60-inch telescope on Mount Wilson.

For more than four decades, the 200-inch (5-meter) Hale Telescope was the largest telescope in the United States. It maintained its supremacy because it is very close to the size limit for telescopes made with a single rigid piece of glass. When the 200-inch telescope was built, rigidity of the mirror was a key requirement if stellar images were to remain sharp as the telescope was pointed at different parts of the sky. Observations of faint objects often take several hours to complete. In order to make such a long observation, the telescope must closely track the object being measured as it rises and sets. As the telescope moves, its primary mirror will tend to sag due to its own weight, and the amount of sag depends on the orientation of the telescope. If the telescope is pointed near the horizon, the mirror is nearly vertical, and the sag will be different from when the telescope is pointed directly overhead, and the mirror is nearly horizontal.

During most of the twentieth century, engineers attempted to minimize the sag by building thick, stiff mirrors and then coupling the mirror to an even stiffer steel mirror cell. For example, if the 200-inch mirror were a solid piece of glass, it would weigh 40 tons. In fact, the Palomar mirror is not a solid piece of glass. Instead, the back of the mirror has a honeycomb structure, much like an egg carton. There are 36 pockets where there is no glass, and these pockets are separated by ribs of glass that help to make the mirror stiff. Even so, this 14.5-ton lightweight mirror, its mirror cell, and the steel mount to support them add up to 530 tons. The telescope is so finely balanced that this massive structure can be moved with a 2-horsepower motor.

Only one telescope was built with an even larger rigid mirror—a 6-meter Russian telescope located in the Caucasus. This mirror weighs 40 tons, and the moving weight of the steel mount is 650 tons. This telescope has never performed as well as the 200-inch telescope, and its massive size is the fundamental problem. As we saw in the previous chapter, turbulence in the atmosphere can blur the images of astronomical objects when they are viewed through a telescope. Turbulence can also be caused if the temperature of the interior of the dome and the structures within it is different from the outside temperature. It takes a long time to adjust the temperature of 650 tons of steel and 40 tons of glass to whatever the outside temperature happens to be. The 6-meter primary mirror adjusts its temperature so slowly that it can tolerate only a 4 degree Fahrenheit temperature change *per day* and still retain the shape required to produce sharp images. The dome is air-conditioned during the day to help keep the interior and the mirror at the likely nighttime temperature, but temperatures can change abruptly in the mountains.

Breaking the Barrier

It was obvious to astronomers that if even larger telescopes were to be built, an entirely new approach would be required. Specifically, engineers had to figure out how to make use of much lighter-weight mirrors. Since the goal was to achieve mirrors with an even larger diameter, the only option was to make the mirrors thinner, and that meant that inevitably they would no longer be rigid. Therefore, ways would have to be found to control the shape of the mirror in real time as the telescope mount moved to point at different parts of the sky.

The dramatic effect that lightweight mirrors have on telescope design is illustrated by a comparison of two 4-meter-class telescopes. The image on the left shows the 4-meter Mayall Telescope on Kitt Peak. The primary mirror in this telescope weighs 15 tons, and the weight of the steel structure that supports its movement is 375 tons. The new Discovery Channel Telescope (DCT) at Lowell Observatory, shown in the right-hand image, has an even larger mirror of 4.3 meters (170-inch diameter), but the mirror is only 4 inches thick, weighs only 3.35 tons, and requires only 135 tons of supporting steel. Compare the massive yoke of the Mayall Telescope with the simple mount of the DCT. Less material means lower cost. Lower mass also means that it is easier to match the temperature in the dome to the outside temperature, minimizing the turbulence that causes bad seeing.

Left: © Darryl Willmarth

Right: Lowell Observatory Archive

We have already noted that the shape of the primary mirror is affected by gravity. The shape of the primary mirror, especially one that is not very rigid, can also be distorted by wind blowing on it and by temperature changes. Domes are designed to shield the telescope as much as possible from the wind, but complete shielding is not possible. Obviously, an opening is required so that the telescope can see the sky.

In addition, modern domes are very well ventilated so that the nighttime air can flow through them and force the temperature of the telescope and the interior of the dome to match the outside temperature and minimize air turbulence. One of the major challenges in designing mirrors larger than 200 inches was to develop control systems that could correct whatever distortions in mirror shape do occur and to do so in real time.

Fortunately, two new technologies came to the rescue. The first was fast computers that could receive information from sensors on the back of the mirror, calculate how the mirror shape was changing, and then send signals to supports behind and at the edges of the mirror. These supports then applied precisely the right amount of force to restore the correct shape. For lightweight, large mirrors, corrections must be made typically about once per second. The second new technology was the development of finite element analysis, a computational tool that allows engineers to calculate how a mechanical structure like a telescope mount or a mirror will behave—before the structure is built. That made it possible to test potential designs in advance and, for example, to use no more steel than absolutely required, thereby minimizing the mass whose temperature had to be made to match the outside temperature at night.

Astronomers and engineers, given the challenge of figuring out how to manufacture and control the shape of very large telescope mirrors, worked out three very different solutions. Before these designs were actually built, there was vigorous debate about which approach would prove to be the best. In the end, advocates of each type of mirror went on a telescope-building spree. Between 1993 and 2004, astronomers in the Southwest and California completed construction of nine telescopes with apertures between 6.5 and 10 meters (250 to 400 inches). As it turned out, all three types of mirrors perform superbly.

The Keck Telescopes

The first large telescopes to be completed were the twin telescopes Keck 1 (1993) and Keck 2 (1996), built by a partnership of the University of California and Caltech and both located on Mauna Kea in Hawaii. Each of the primary mirrors is 10 meters (394 inches) in diameter and is composed of 36 individual hexagonal segments. Each segment is 1.8 meters wide (70 inches) and about 3 inches thick, and each weighs about half a ton. The segments are closely packed, with their edges separated by 0.12 inches. The total weight of

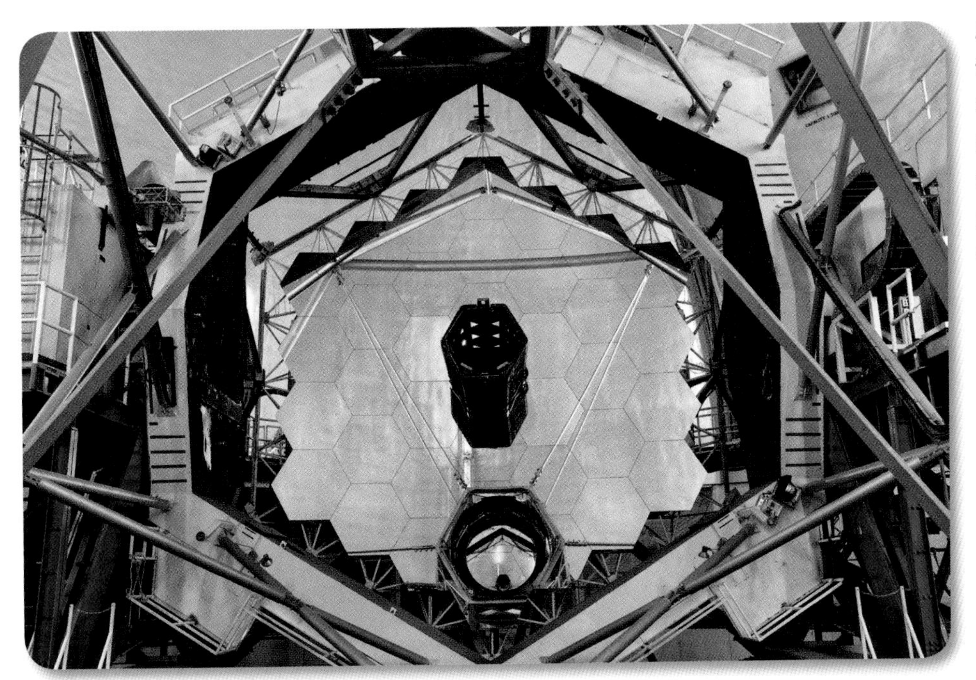

A bird's-eye view of the Keck primary mirror. Note the hexagonal segments of glass that together make up the primary mirror.
Keck Observatory

all 36 segments is about the same as the weight of the 200-inch Hale Telescope, but Keck's total light-collecting area is four times greater, which means it can see objects four times fainter in the same amount of observing time.

The big challenge with this design is to force the segments to act together as if the mirror were made of a single piece of glass. To solve this challenge, each mirror segment is equipped with sensors that measure its position relative to neighboring segments. This information is instantly and continuously sent to a computer. If a segment is out of position by as little as a millionth of an inch (a human hair is about 1,000 times thicker), then the computer calculates what adjustment is required and sends the information to adjusting mechanisms on the back of the mirror. This whole process can be completed in half a second, and then the process begins again. These adjustments keep the mirrors aligned so that they produce a single sharp image as the telescope moves around the sky.

The Hobby-Eberly Telescope (HET)

The University of Texas also built a telescope with segmented mirrors but adopted a novel, very clever, and low-cost approach. The primary mirror of the HET is 11.1 × 9.8 meters and is the largest one built so far. The mirror has 91 segments; each segment is one meter

The Hobby-Eberly Telescope (HET). The primary mirror is made of hexagonal segments like the Keck primary. In the case of the HET, however, the primary remains tilted always at a fixed angle and rotates only in azimuth (i.e., from north to south). In order to observe at different altitudes, the instrument that detects light moves along the black truss mounted at the top end of the telescope.

EXPLORING ON YOUR OWN

Choosing a Telescope

By the end of this chapter, you may be starting to think about acquiring a telescope of your own. Type "choosing a telescope" into a search engine, and you will find lots of advice. Look at some of the noncommercial sites like *Sky & Telescope* first. You will have to make the same choices that astronomers have had to make. How big should your telescope be? A larger primary has more light-gathering power and will make it possible to see fainter objects, but for amateurs as well as professionals, larger telescopes usually cost more. There is also the question of portability. Just as photographers say that the best camera is the one you have with you, the best telescope is one that you can easily take out to your backyard or transport to a dark site. Do you want a refractor or a reflector? A refractor is usually more rugged but more costly and typically about four feet long. Reflectors are usually more compact and give you the most light-gathering power for the money. The mount is critical. An optically superb telescope on an unstable mount is virtually useless.

Be warned that no telescope will allow you to see images like the glorious pictures that Hubble has given us. Those images are made with long exposures through different filters, and these separate images are combined later with an image-processing program, such as Photoshop. The eye sees an image instantaneously—in effect, a very short exposure—and the eye is insensitive to color at low light levels.

Because a telescope is a significant investment, the best advice is to do your homework first. Even the computer-controlled telescopes that will go automatically to the object of your choice are easier to use if you are already familiar with the sky. Learn the constellations. Identify the planets. Binoculars, either on a stable mount or image-stabilized, will allow you to see the colors of stars; red stars are the coolest, blue-white stars are the hottest, and yellow stars are somewhere between these extremes. With binoculars you can also observe many double stars, star clusters, Jupiter's four brightest moons, and some of the mountains and craters on our own Moon. Lunar features are most easily seen along the terminator—the boundary between the sunlit dayside and the dark nightside.

See if there is an astronomy club nearby. Participate in stargazing parties and find out what kinds of telescopes these dedicated amateurs prefer and why. All of this information will help you to decide whether you do want a telescope, and if so, which one is the best choice for the way you plan to use it. ●

(39 inches) across and weighs about 250 pounds. What makes the HET design revolutionary is that the primary mirror always points at a fixed elevation of 55 degrees above the horizon. The primary does rotate in azimuth, that is, it can point in any direction from north to south. The advantage of this arrangement is that the mirror does not change its direction relative to gravity, and the problems of supporting it are greatly simplified. The HET cost only 15–20 percent as much as a fully steerable telescope of comparable aperture.

In order to make a long exposure observation, it is necessary to track a star as it moves across the sky. The HET does this by moving the instrument used to record the observations rather than moving the primary mirror in altitude as well as azimuth. It is as if an eyepiece rather than the entire telescope is tracking the star. The instrument is mounted

13 meters above the primary mirror. Because the instrument changes its position relative to the primary mirror, it can see only a portion of the primary's surface at any given time. Think of what happens when you move your eye relative to the eyepiece of binoculars—some of your view is blocked. The consequence of the HET design is that the instrument can only see an amount of mirror that is equivalent to a conventional telescope with a diameter of 9.2 meters. Accordingly, when we compare the light-gathering power of the HET with other telescopes, we say it has an effective aperture of 9.2 meters.

The Gemini Telescopes

The most obvious option for making a very large telescope mirror is simply to make one out of a single monolithic mirror that is very thin and has a large diameter; such a thin mirror is called a meniscus. This solution eliminates the challenge of controlling 36 individual mirrors but instead requires control of the shape of a mirror that is very flexible. Meniscus mirrors were used in the twin Gemini Telescopes, which were built on Mauna

Inspection of the Gemini North mirror. The mirror is only 8 inches thick. Note the person for scale in the center of the mirror.

Gemini Observatory/AURA

Kea in Hawaii and on Cerro Pachon in Chile by an international consortium that initially involved the United Kingdom, Canada, Chile, Argentina, and Brazil, as well as the United States. The project was initiated by the National Optical Astronomy Observatory and carried out by a team based in Tucson. (The European Southern Observatory also adopted meniscus mirrors for four 8.2-meter telescopes that were built in Chile.)

A mirror blank is one that is not yet polished to the shape required to bring light to a sharp focus. The Gemini mirror blanks were manufactured by Corning, the same company that made the 200-inch blank for the Hale Telescope on Palomar. Each Gemini blank is 318 inches (8.1 meters) in diameter but only 8 inches thick. To form each blank, 55 hexagonal blocks of low expansion glass were fused together at a temperature of 1,700 degrees Celsius. Imagine the Gemini blank as being like the Keck mirror, but with the individual hexagonal segments fused together.

After fusing, the Gemini mirrors were flat. Each was then placed in an oven on top of a mold that had a shape close to the one that would be required in the telescope. The temperature was then raised just enough to cause the mirror to slump down over the mold.

A fish-eye view of the Gemini North interior. A laser beam through the open dome slit is being used for adaptive optics observations. The openings to the sky that circle the dome are for ventilation. The flow of outside air through the dome helps to ensure that the temperature inside matches the temperature outside. The goal is to minimize turbulence that would produce bad seeing.

Gemini Observatory/AURA

After slumping, the mirror blanks were shipped to REOSC Optique in Paris, France, where they were polished to the required final shape, which is accurate to better than a millionth of an inch.

Such large, thin mirrors are, by the standards required for astronomy, quite flexible. Some skeptical astronomers even compared them with a sheet of paper flapping in the wind. To preserve the perfect shape of the Gemini mirrors, 120 actuators apply forces to the back of the mirror to push it back into the correct shape as the telescope tracks across the sky. The Gemini North Telescope in Hawaii began observations in 1999, with Gemini South in Chile following two years later.

Steward Observatory Mirror Laboratory

Both Keck and Gemini solved the problem of controlling the shapes of their mirrors, and both types of mirrors produce superb images. The astronomers at Steward Observatory of the University of Arizona, however, decided to minimize the control problem by building a stiffer but very light-weighted mirror—and to manufacture the mirrors at the university itself rather than seeking a commercial manufacturer as both Keck and Gemini did.

How do you transport a large mirror? Very carefully! An 8.4-meter mirror from Steward weighs about 20 tons; the mirror cell weighs another 10–15 tons. The combination is much too heavy to be lifted by helicopter. The mirror must be transported by truck to the observatory site or to the port from where it will be shipped overseas. This image shows the Large Binocular Telescope (LBT) mirror and cell being transported up Mount Graham. The transporter has been designed so that the mirror can be transported either flat or tilted if the road is narrow. The 8.4-meter size of this mirror is about the largest that can be transported over U.S. roads.

LBT Observatory

The Steward mirrors are made of borosilicate glass, which is the kind of glass used to make cookware (e.g., Pyrex). The mirror consists of two main parts. The front surface, which is polished to the precise shape needed to produce images, is only a little over an inch thick (remember the Gemini mirrors are about 8 inches thick). To add rigidity, this faceplate is backed by a honeycomb structure, also made of borosilicate. The walls of the honeycomb are typically about an inch thick, and the walls are separated by a distance of about 8 inches. Most of the back of the mirror is empty space! To see why this works, cut a piece of paper the size of the bottom half of an egg carton. Then hold the paper by one corner and see how much it bends if you tilt it or even just blow on it. Now attach the paper to the egg carton, and see how much stiffer it is. While the Steward mirrors are more rigid than the alternatives, active, real-time control is still required to preserve the image quality under observing conditions.

The twin 8.4-meter mirrors of the LBT. Note the two people in the cherry picker near the right-hand side of the image to get a sense of the scale of the telescope. The men are cleaning the mirror.

How to Make a Really Big Mirror

One of the mirrors made by Steward Observatory is destined for the Large Synoptic Survey Telescope. The LSST is an 8.4-meter telescope with a unique design. It will have a much larger field of view than other telescopes of comparable size and has been designed to survey the sky very rapidly. Specifically, the plan is to complete two full scans of the entire visible sky every week, and to do this over and over for 10 years. After 10 years, we will have 800 images measured at wavelengths from the ultraviolet to the near infrared of every patch of the southern sky visible from the chosen site in Chile near the Gemini South Telescope.

The resulting "movie" of the sky will be used for literally hundreds of research projects. One of the prime motivations for building the telescope is to measure the properties of the mysterious dark matter and dark energy described in the final chapter of this book (page 127). The many repeated observations will be well suited for detecting objects that change position, such as potentially hazardous asteroids that might impact the Earth (see Chapter 4). Variations in brightness will be detected, and so the LSST will be ideally suited for discovering supernovae, which are the explosions of dying stars, and other transient events.

The LSST project is being carried out jointly by the National Science Foundation and the Department of Energy, and the telescope will be located in the foothills of the Andes in Chile. The project team is based at the National Optical Astronomy Observatory in Tucson. First light is scheduled for 2019 and will be followed by two years of commissioning and testing. Scientific observations will begin two years after first light. ●

▲ **An artist's conception of the enclosure for the LSST telescope.** Note the large openings for ventilation.

LSST Project/AURA/NSF

These images show how the LSST mirror was made.

▲ **Chunks of glass to be used to make the LSST mirror.** The mirror requires nearly 52,000 pounds of E-6 low expansion glass, supplied by the Ohara Corporation in Japan. Each piece is unpacked, weighed, and graded for proper placement in the oven mold. Ray Bertram, Steward Observatory Mirror Lab

▲ **Creating the mold for the LSST mirror.** Each hexagonal column is 20 centimeters in diameter and 90–100 centimeters high. The space between the columns is 2.5 centimeters. The LSST mirror will be 8.4 meters in diameter, and 1,700 columns are required. The columns have a higher melting temperature than glass.

V. Krabbendam, LSST Corporation

▲ **The glass chunks are loaded into the mold.** An 8.4-meter mirror requires about 20 tons of glass. The glass will be heated to a little over 2,100 degrees Fahrenheit. At this temperature, the glass is about the consistency of honey and will flow down into the spaces between the columns in the mold. Ray Bertram, Steward Observatory Mirror Lab

▲ **The oven used to melt the glass.** The glass is heated in an oven that spins at about 7 rpm so that centrifugal forces cause the molten glass to take on a parabolic shape that is close to the same shape as required in the telescope. The glass is heated for about seven days, and cooled very slowly for about eleven weeks. This image shows the oven spinning up to make the fourth mirror for the Giant Magellan Telescope (see end of this chapter). Ray Bertram, Steward Observatory Mirror Lab

▲ **The LSST mirror after it was removed from the oven.** The LSST team surrounds the mirror. Your author served as director of this project during the design phase and appears in turquoise in the front row on the right. Howard Lester, LSST Corporation

▲ **Polishing of the LSST mirror.** After the mirror was cast, the staff at the Steward Mirror Laboratory undertook the challenging task of polishing the mirror to the exact shape required to produce exquisite images in the telescope. This polishing process must achieve an accuracy of better than one-millionth of an inch. The entire process from casting to a completed, polished mirror takes about four years. LSST requires three mirrors to achieve the wide field of view necessary to survey the sky rapidly and is unique in having the first and third mirrors fused into the single piece of glass shown above. Note that this mirror has two distinct parts, with the center part somewhat deeper than the outer annulus. Light from the sky falls first on the outer annulus of the mirror, is reflected up to a second mirror (not shown), and then is reflected down onto the inner, deeper part of this mirror, which directs the light to a focus at the camera. E. Acosta, LSST Corporation

EXPLORING ON YOUR OWN

Very Large Telescopes

Very large telescopes are very expensive to build. The planned 30-meter-class telescopes are projected to cost over a billion dollars each. Given this big investment, astronomers pick the very best sites for them. These sites are often remote, difficult to reach, and may not be open to visitors.

Astronomers are, however, committed to sharing the excitement of their new projects and discoveries through the Web. Type "ESO webcam" into a browser, and you will find live Web feeds from several ESO sites. On the LSST website, you will find a webcam with live views of the construction site in Chile. Sometimes there is no apparent action for days, but at other times there will be visible activity. All of the large telescope projects—Giant Magellan Telescope (GMT), Thirty-Meter Telescope (TMT), and the ESO Extremely Large Telescope (E-ELT) have websites for the public with image galleries and information about progress. TMT has a twitter feed, and the other projects are increasing their presence on social media. There are many videos on YouTube about these large telescope projects.

When ongoing work on large mirrors permits, the Steward Observatory Mirror Laboratory offers tours of the mirror casting and polishing facilities. Check their website for availability and times.

Seeing a large telescope for the first time is awe-inspiring. Three large telescopes in the Southwest offer daytime tours for the public. These are the 9.2-meter Hobby-Eberly Telescope (HET) at McDonald in Texas, the Large Binocular Telescope (LBT) on Mount Graham in eastern Arizona, and the 6.5-meter MMT, a short drive south of Tucson. Some are open only on weekends or during the summer months. Check their websites for times and availability in advance of your visit. Also imposing are the 4-meter Mayall Telescope at Kitt Peak and the 200-inch telescope on Palomar. ●

So far, the Steward Observatory Mirror Lab has made 18 mirrors and more are planned. The mirrors were initially small—one that was made in 1985 and is now in the Vatican telescope on Mount Graham, is only about 70 inches in diameter. Over time, as Steward staff became more experienced with the techniques, mirrors grew, and the largest mirrors made there are 330 inches (8.4 meters) in diameter. Two of these mirrors are in the Large Binocular Telescope on Mount Graham. This telescope can be used not only as a conventional telescope to observe faint objects but also as an interferometer.

Giant Telescopes for the Twenty-First Century

Given access to space telescopes, why do astronomers continue to build ground-based telescopes? Telescopes in space have revolutionized astronomy, but ground-based telescopes continue to be critical for advancing our understanding of the Universe. Because ground-based telescopes are less costly to build, they can be made much larger than space telescopes, and we can have more of them. This means that many different research programs can be pursued at the same time, and progress is more rapid. The James Webb Space Telescope is the successor to Hubble and has an aperture of 6.5 meters. It is scheduled for launch in 2018 at a cost of nearly 9 billion dollars. The next generation of ground-based

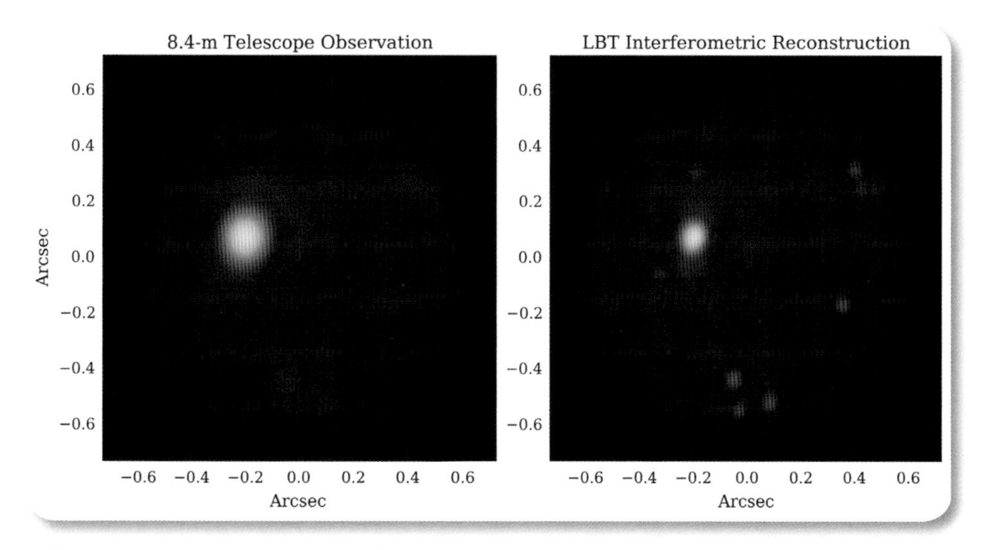

Volcanoes on Jupiter's moon Io. Left: A simulation of the way Io would appear if it were observed with just one of the LBT mirrors. Right: An image made by using the LBT as an interferometer; light from both mirrors is combined to make a single image that clearly shows several eruptions. When the two light beams are combined, the LBT has a resolution equivalent to that of a single mirror with a diameter of 23 meters. LBTO

telescopes will have apertures of 22 to 39 meters and will cost 1–2 billion dollars. Bigger apertures are useful for observing extremely faint sources and also for spectroscopy (see Chapter 5, page 98). The Hubble Telescope, for example, has returned fantastic images, but spectroscopy is often required to understand those images. It takes a larger aperture (i.e., a larger primary mirror) to observe the spread-out light of a spectrum than it does to take an image through, for example, a red filter.

Each new view of the Universe with increasingly powerful telescopes just makes astronomers want to see more—to explore what is just beyond the limits of what they can already see. There are now plans underway to break the 10-meter size barrier. Three telescopes have been designed that range in size from 22 meters to 39 meters. Because an 8-meter-class mirror is about the largest size that can be transported from the factory where it is made to its final destination on a mountaintop, these three giant telescopes will all be made by combining smaller mirrors on a single mount.

The projects to build these telescopes all have plans to address the biggest unanswered questions about the origin and evolution of the Universe and of life itself.

> *Is there life elsewhere in the Universe? Or are we alone?* The giant telescopes will be able to image Earth-like planets around nearby stars and analyze their atmospheres to look for molecules that are indicators of biological activity—life.

How and when did galaxies form? How do they change with time? When did the first stars form? When did larger groups of stars gather together to form galaxies? Galaxies collide and merge to form still larger galaxies. How important are mergers in building the largest galaxies? How does the merger rate vary with time? When and where were elements heavier than hydrogen and helium formed?

What is the role of black holes in shaping the evolution of galaxies? Black holes with masses of a million to more than a billion times the mass of the Sun lie at the centers of galaxies. The most massive galaxies have the largest black holes. Why? And how do black holes and galaxies affect the evolution of each other?

The 8- and 10-meter telescopes now in operation, along with the Hubble Telescope, have brought astronomers tantalizingly close to observing directly the formation of the first generation of stars and galaxies (see Chapter 6). It is amazing to think that, with the next generation of telescopes, we may soon be able to see directly—not just imagine—how some of the very earliest stages in the history of the Universe unfolded.

The Giant Magellan Telescope (GMT)

The GMT will have a segmented mirror telescope made up of six 8.4-meter mirrors that surround a central mirror, also 8.4 meters in diameter. The mirrors are all being made by the Steward Observatory Mirror Laboratory. The resulting optical surface will be 80 feet in diameter. The telescope will be built at the Las Campanas Observatory, which is on the southern edge of the Atacama desert in Chile. The project is being carried out by an international consortium involving several countries and U.S. universities.

Giant Magellan Telescope Organization

TMT Corporation

The Thirty-Meter Telescope (TMT)

The primary mirror of the TMT Observatory will be a segmented mirror similar to the mirrors in the Keck telescopes. The mirror will be 30 meters in diameter and will consist of 492 segments. The telescope will be located on Mauna Kea, Hawaii, and the project, like the GMT, is being carried out by an international consortium that includes several countries as well as the University of California and Caltech.

The European Extremely Large Telescope (E-ELT)

When completed, the E-ELT will be the largest of the planned telescopes. Its mirror will be 39 meters in diameter, which is nearly half the length (42 yards) of a football field. The primary mirror will consist of 798 hexagonal segments, each of which is 1.4 meters (4.6 feet) across but

ESO

only about 2 inches thick. The E-ELT will be located in the Atacama desert on Cerro Armazones, a 10,000 foot peak in the coastal mountain range about 80 miles from Antofagasta, Chile. In this image, an early design of the telescope enclosure is compared with the size of the great pyramids. The domes of the four 8.2-meter VLT instruments at the European South Observatory (ESO) are also shown for comparison. This is an artist's conception. The E-ELT and the VLT telescopes will be on different mountains—and, of course, the pyramids are in Egypt.

The James Webb Space Telescope (JWST)

Although this chapter has focused on the development of ever-larger ground-based telescopes, technology for space-based telescopes is also advancing dramatically. Scheduled for launch in late 2018, the JWST will be the next major space telescope. Optimized for infrared observations, this telescope will have a mirror that is 6.5 meters in diameter. For comparison, the Hubble Space Telescope mirror is only 2.4 meters across. The instruments on JWST will make it possible to study the entire history of the Universe from the formation of the first galaxies to the characteristics of solar systems with Earth-like planets that might be capable of supporting life.

Follow the progress of JWST at www.jwst.nasa.gov.

An artist's conception of the James Webb Space Telescope. The 6.5-meter primary mirror is made of eighteen separate beryllium mirrors, which will unfold and adjust to the correct shape when the telescope reaches orbit. The JWST is also equipped with a shield, which is the size of a tennis court and will protect the telescope and instruments from the heat of the Sun. Northrop Grumman

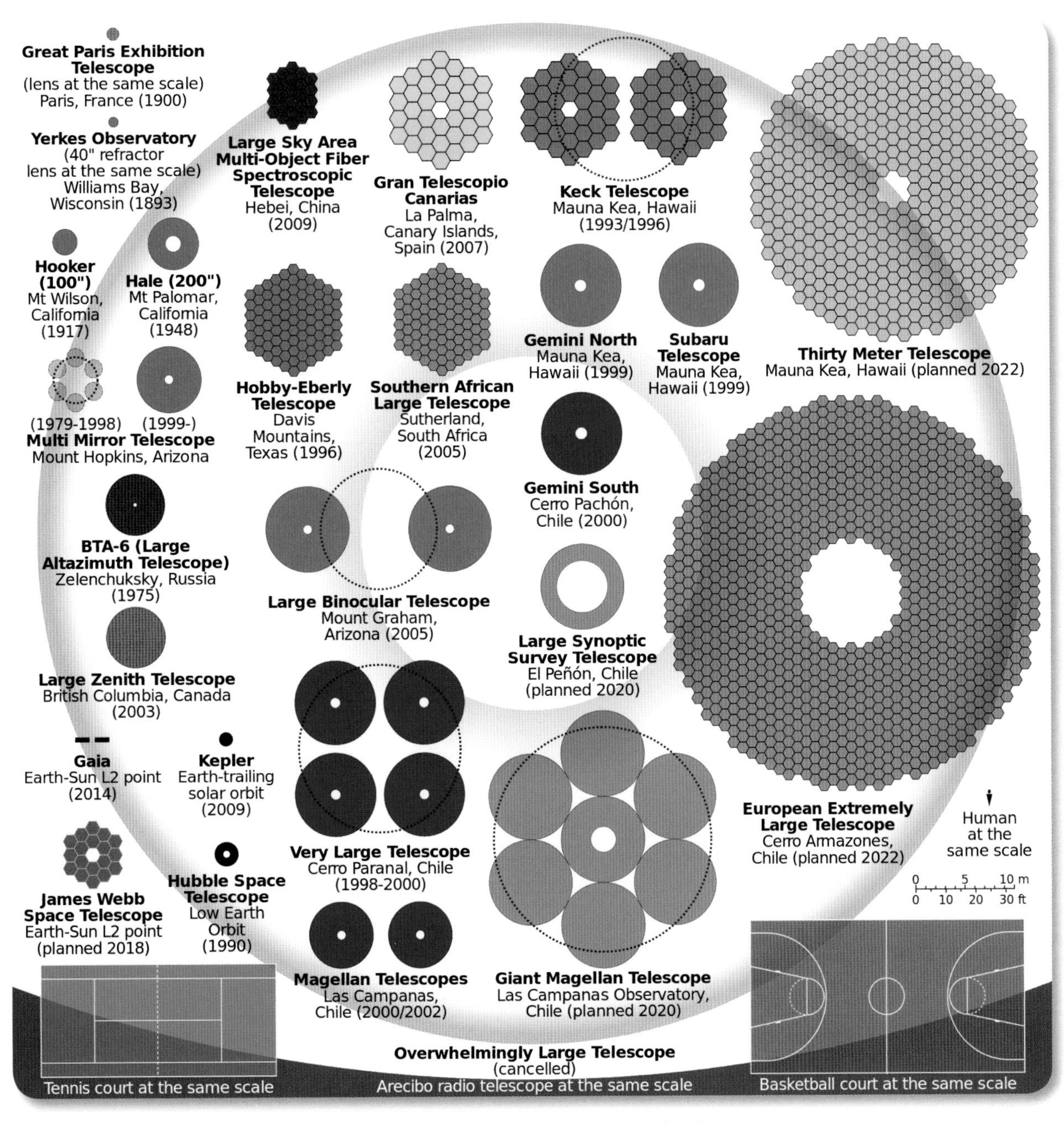

Comparison of sizes of the primary mirrors for some notable telescopes. This diagram summarizes the history of telescope building during the twentieth century and shows telescopes planned for completion in the early twenty-first century. Telescope mirrors shown in blue are in the United States. Telescopes in green are in Chile. Telescopes in black are in space. Lighter shades of these three colors show telescopes currently planned or under construction. Cmglee/Wikipedia

A scaled drawing of the solar system. The planets in the upper part of the diagram are shown in order of their distance from the Sun and in their correct relative sizes. In order, these planets are Mercury, Venus, Earth, Mars, Jupiter, Saturn, Uranus, Neptune, and dwarf planet Pluto. In the lower part of the image, the orbits of these planets are shown with their correct relative sizes. The sizes of the planets are greatly exaggerated relative to the sizes of their orbits. NASA/JPL

4 The Solar System

Astronomical observatories in the Southwest, California, Texas, and Hawaii have all played major roles in exploring the solar system. Examples include:

- The discovery of Pluto, the most distant planet in the solar system, and the discovery a century later of dwarf planets, which led to the demotion of Pluto to dwarf planet status;
- The discovery of Comet Shoemaker-Levy, which broke apart and crashed into Jupiter, producing fireballs and leaving behind dark debris spots that were visible for months;
- Systematic searches for potentially hazardous asteroids that might crash into the Earth and cause significant damage;
- Observations of clouds on Uranus and storms on Saturn's moon, Titan;
- Monitoring of the Sun to try to improve models for predicting space weather.

Observations from the surface of the Earth, however, can only hint at the beauty and complexity of our solar system. Carl Sagan once told an audience of students that this was a very special time to be alive. In the mid-twentieth century, all that we knew about the planets and their moons was based on images shrunk by distance and blurred by our own atmosphere. In the lifetimes of those students, Sagan pointed out, spacecraft would visit all of the planets in our solar system, and close-up observations would revolutionize our knowledge of these fascinating and diverse worlds.

Sagan himself lived to see images taken by spacecraft passing near all of the planets except Pluto. The New Horizons spacecraft finally reached the most distant planet (now relabeled a dwarf planet) in our solar system in 2015, nearly 20 years after Sagan's death in 1996. Each planet and many of their moons are now distinct worlds with unique, and in many cases completely unexpected, characteristics. Most of the spacecraft that made these journeys deep into interplanetary space were built at the Jet Propulsion Laboratory in Pasadena.

Mercury

Fun Facts

- *Mercury is the smallest planet. It is 3,000 miles in diameter or about 1.5 times the diameter of Earth's Moon.*
- *Mercury is the planet closest to the Sun. Its average distance from the Sun is 36 million miles.*
- *Of all the planets, Mercury has the biggest range in temperature: From 800 degrees Fahrenheit in the daytime to –290 degrees at night.*

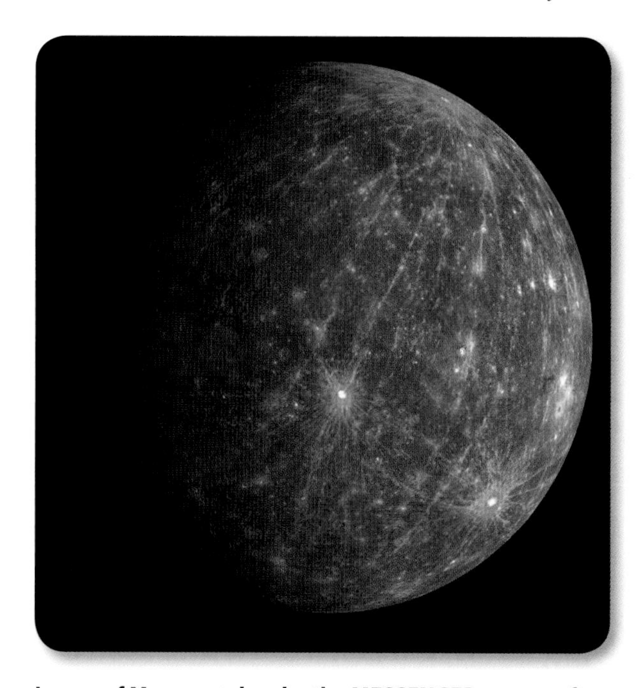

Image of Mercury taken by the MESSENGER spacecraft. The surface of Mercury is similar to that of Earth's Moon, with the most prominent features being impact craters produced by collisions with asteroids, which are small, airless rocky bodies that orbit the Sun. The bright crater just south of the center of the image is Kuiper, named after Arizona planetary astronomer Gerard Kuiper. A large pattern of rays extends from a relatively young crater in the northern part of Mercury to regions south of Kuiper. There is another rayed crater southeast of Kuiper, near the limb (edge) of the planet. NASA/Johns Hopkins University Applied Physics Laboratory/Carnegie Institution of Washington

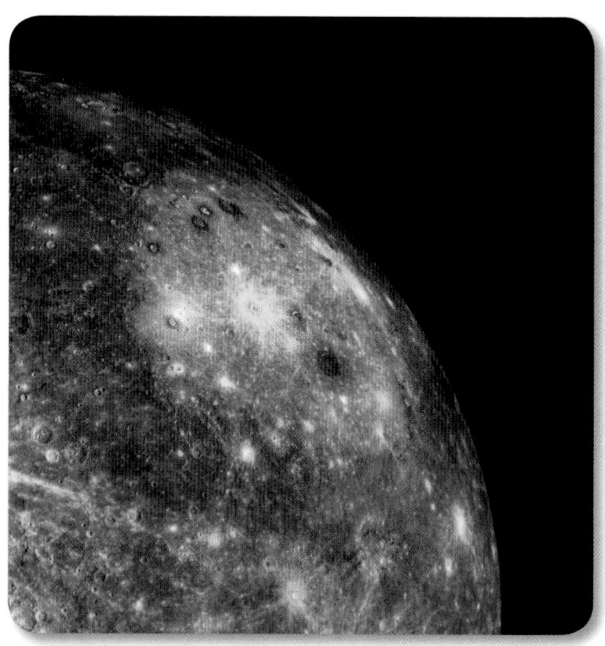

The Caloris Basin on Mercury. This basin was created during the early history of the solar system by the impact of a large asteroid-sized body. Almost 1,000 miles in diameter, Caloris is one of largest impact basins in the solar system. Orange splotches around the basin's perimeter are thought to be ancient volcanic vents. This false-color image accentuates subtle color differences in order to detect differences in geologic features and chemical composition. NASA

An example of true-color (left) and false-color (right) images. These are photos of the wall in the entrance tunnel of the Queen Mine in Bisbee, Arizona. The left-hand image is what the eye sees. In the right-hand image, colors have been enhanced with a photo-editing program. This enhancement of color makes it much easier to see that there are different minerals in the wall of the mine. This same technique has been used to make the false-color image of Mercury. © Sidney Wolff

Mercury looks much like the Earth's Moon. The largest craters on its pock-marked surface were formed by impacts from asteroids early in the history of the solar system about 4 billion years ago. Collisions with smaller rocky meteoroids since then have formed the many smaller craters that dot the surface. Mercury's atmosphere is so thin that it hardly deserves to be called an atmosphere. The atoms that form it were blasted off of Mercury's surface by meteoroid impacts and by the solar wind, which consists mainly of protons and electrons ejected from the Sun at speeds that can reach a million miles per hour. These atoms do not stick around long enough to form a permanent atmosphere but are driven away from Mercury by solar radiation pressure—and are replenished by new impacts and blasts from the solar wind.

Venus

Fun Facts

- *Venus is the planet that comes closest to Earth. It is the brightest of the planets that can be seen with the unaided eye and appears even brighter than Sirius, which is the brightest star.*
- *A day on Venus is longer than its year. It takes Venus 243 Earth days to rotate once on its axis but only 225 Earth days to go all the way around the Sun and complete a Venus year.*

Maat Mons on Venus as detected by radar measurements. Maat Mons rises almost 3 miles above the surrounding plain, which is covered for hundreds of miles by lava flows. Maat was an Egyptian goddess of truth and justice. JPL/NASA Planetary Photojournal

- *Spacecraft have landed on the surface of Venus but none has survived more than a couple of hours because of high temperatures.*
- *All but three of the surface features on Venus are named for famous women or goddesses from various cultures. Maxwell Montes is the only feature on Venus named after a man—James Clerk Maxwell, a British scientist who developed equations that describe the behavior of electricity, magnetism, and light. The other two exceptions are Alpha and Beta Regio. These three were named before the convention of using female names was adopted.*

With a name like Venus, the planet that is second closest to the Sun sounds like an alluring place to visit. It is about the same size as the Earth, and like the Earth, it has clouds in its atmosphere. For a long time, scientists thought it would be the planet most similar to Earth. Sagan once imagined that it was a livable planet with swamps and oceans. The clouds on Venus are so thick, however, that scientists could not observe the surface. These clouds hid the awful truth—now, the adjective most often used to describe Venus is "hellish."

Spacecraft measurements have been responsible for turning Venus from an imagined Eden to a hell. Although visible light cannot penetrate the opaque clouds, radar and infrared energy can. Space missions to Venus have shown that the surface temperature is 900 degrees Fahrenheit and is hot enough to melt lead. The atmosphere is composed mainly of carbon dioxide with clouds composed of droplets of sulfuric acid. Atmospheric pressure is about 90 times higher than at the surface of the Earth or about equivalent to the pressure in the Earth's oceans at a depth of 0.6 miles. Radar maps made by orbiting spacecraft show that Venus has more than 1,000 volcanoes. About 80 percent of the surface of Venus is covered by lava flows. It is not clear whether volcanoes are still active, but researchers have concluded that Venus was completely covered by lava flows that occurred 300–500 million years ago. There are only trace amounts of water and oxygen in the atmosphere.

Why is Venus so different from Earth? Scientists think that the two planets may have been similar up to about 600 million years after they first formed about 4.5 billion years ago. Venus may have had oceans of water, and the Earth may have had a thick carbon dioxide atmosphere and even sulfuric acid clouds. Unlike Venus, the Earth was cool enough to retain its oceans, and water is the key to controlling carbon dioxide, which is a greenhouse gas. A greenhouse gas absorbs heat emitted from the surface of a planet and then reemits it, sending it back downward where it can heat the surface still more. Earth's oceans absorb a lot of carbon dioxide, keeping it out of the atmosphere. Also, water and carbon dioxide react with silicates to lock up carbon dioxide in rock.

Venus is a little closer to the Sun and has always been at least a little hotter than the Earth—just hot enough billions of years ago to evaporate any surface water that it may have had initially. Water vapor in the atmosphere, like carbon dioxide, is a greenhouse gas. As the surface of Venus got even hotter because of this greenhouse effect, water evaporated ever more quickly. More water in the atmosphere trapped more heat, and Venus got hotter still—a runaway effect that evaporated all the water. Ultraviolet energy from the Sun and solar wind particles have enough energy to break up water molecules in the Venus atmosphere into hydrogen and oxygen. The lighter hydrogen then escaped into space, and without hydrogen, water was permanently lost from Venus. Without water, there was no way to extract carbon dioxide from the atmosphere.

It thus seems likely that a rather small difference in temperature billions of years ago allowed life to develop and flourish on Earth while rendering Venus completely uninhabitable.

Which image is Mars and which is the American Southwest? In the left image, we see dramatic buttes and layers on the lower slopes of Mount Shart on Mars. Colors have been adjusted to show the rocks as they would appear under daytime lighting conditions on Earth. The right image was taken near the border of Arizona and Utah. The real clue is the deep blue sky, which is produced by the scattering of sunlight in the Earth's atmosphere. The thin Martian atmosphere does not scatter sunlight as effectively. Left image: NASA/JPL-Caltech/MSSS; right image: © Sidney Wolff

Mars

Fun Facts

- *Olympus Mons on Mars is the largest volcano in the solar system. It is about the size of Arizona (its diameter is about 375 miles), and it rises nearly 9 miles above the surrounding plains.*
- *Most of the eruptions that created Olympus Mons probably happened billions of years ago. However, there are only a few impact craters on its flanks, which suggests that as recently as 25 million years ago, a few lava flows may have covered up older impact craters.*
- *Mars also has the largest canyon in the solar system. Valles Marineris is 2,500 miles long, 125 miles wide in places, and more than 4 miles deep. For comparison, the Grand Canyon is just a little over a mile deep.*

Mars would look very familiar to a visitor from Earth. Like the Earth, Mars has volcanoes and deep canyons, deserts, and polar icecaps. The surface of Mars looks red because, just as in the western United States, iron causes the soil to rust (the technical term is oxidize). Mars even has seasons. However, the similarities end there. Our hypothetical visitor would need a space suit to survive. The Martian atmosphere is mainly carbon dioxide and is very thin. Atmospheric pressure at the surface of Mars is only about one percent of the atmospheric pressure on Earth.

NASA's Curiosity Mars rover and its tracks are visible in this view from orbit, by the High Resolution Imaging Science Experiment (HiRISE) camera on NASA's Mars Reconnaissance Orbiter. The rover is the bright blue object at about a two o'clock position relative to the largest butte in the lower left quadrant of the image. Curiosity entered the area included in this image along the tracks visible near the upper left corner. The distance between parallel wheel tracks is about 9 feet. The area included in the image is about 1,200 feet wide. This is a false-color image with the colors exaggerated to make differences in Mars surface materials more apparent. As a result, Curiosity appears bluer than it really is. NASA/JPL-Caltech/University of Arizona

Mars is too cold for liquid water to survive on the surface for any length of time. While temperatures can reach 70 degrees Fahrenheit at noon on the equator in summer time, temperatures can plunge to –100 degrees Fahrenheit at night. Various Martian spacecraft have, however, found geologic features that suggest that Mars long ago may have had liquid water on its surface. Images of Mars show features similar to flood plains and channels that are seen on Earth and that we know were formed by flowing water.

Recent research has found even stronger evidence that Mars was once very wet indeed. In fact, it may have had an ocean that covered about 20 percent of the Martian surface and that was probably up to a mile deep in some locations. To arrive at this conclusion, scientists measured the ratio of ordinary water, H_2O, to heavy water, or HDO. Ordinary water consists of two protons and one oxygen atom, but in heavy water, one proton is replaced by deuterium, which has one neutron and one proton. Because H_2O weighs less, it is much easier for it to evaporate into the atmosphere and escape into space, leaving Mars forever. The scientists found that Mars has much more HDO relative to H_2O than is found on Earth, which implies that Mars has lost a lot of ordinary water. Quantitatively, the scientists concluded that 87 percent of the water once found on Mars must have been lost to space. Some of the remaining water on Mars is found in the frozen polar ice caps. There is also evidence for water ice glaciers between latitudes 30° and 50° in both the northern and southern hemispheres. The glaciers are covered by a thick layer of dust and cannot be seen in images taken in visible light. Recently, scientists have imaged dark streaks on Mars that appear during warm weather and disappear when it gets colder. Analysis indicates that these dark streaks are caused by very salty water dampening the soil. This is the first evidence that small amounts of water still flow on the surface of Mars.

Water is an essential ingredient for life as we know it, so understanding the history and abundance of water is key to understanding whether life ever developed on Mars. Learning more about how much subsurface frozen water remains on Mars may also be key to identifying resources to support human exploration.

The Gas Giants

Fun Facts

- *Jupiter's Giant Red Spot is a storm larger than the diameter of Earth that has persisted for hundreds of years. Since 1979, it has shrunk in size from 14,500 miles across to just over 10,000 miles across. If it continues to shrink at the current rate, it will be gone by 2030.*
- *If you could find a bathtub big enough, Saturn would float. Its density is lower than that of water and lower than any of the other planets.*
- *Uranus's rotation axis is tilted almost parallel to the plane of its orbit. In 1986, when Voyager flew by, the south pole of Uranus was pointed almost directly at the Sun, and it was summer there. The North Pole was experiencing a long dark winter with no sunlight at all for several years. Because it takes 84 years for Uranus to go completely around the Sun, the North Pole won't see summer until 2028, half of a Uranus year later. No one knows why Uranus is tipped on its side, but perhaps sometime in its early history, Uranus was flipped nearly 90 degrees by a collision with another planet-sized body.*
- *Many theorists think that Uranus and Neptune were formed in the neighborhood of Jupiter and Saturn and were then flung farther out into the solar system by gravitational interactions with the two most massive planets.*

Jupiter, Saturn, Uranus, and Neptune are known as the Jovian (Jupiter-like) planets because they are all gigantic compared with Earth. The Jovian planets are also referred to as the gas giants, although some or all of them might have small rocky cores. This diagram shows the approximate relative sizes of the Jovian planets. Lunar and Planetary Institute

We now leave the rocky planets—Mercury, Venus, Earth, and Mars—behind and journey to the gas giant planets of the outer solar system. The four planets closest to the Sun formed where the temperature was so high that only rock-like materials containing elements such as iron, nickel, aluminum, and silicon could form solids that gradually came together under the influence of gravity to build planets. In the outer reaches of the solar system where the temperature was lower, ices composed of carbon, nitrogen, and oxygen could form. Frozen water, methane, and ammonia were common, and rocky material was present as well. Initially, the rocky materials and ices began to come together to build the cores of the giant planets we see today. As these cores grew, they became so massive that they could capture not only more and more ices but also gaseous hydrogen and helium, which are the two most abundant elements in the Universe.

Jupiter is the most massive planet, and both it and Saturn, the second most massive planet, consist mostly of hydrogen and helium. In addition to hydrogen and helium, the interiors of Uranus and Neptune contain large quantities of water, methane, and ammonia.

Jupiter, Saturn, Uranus, and Neptune are similar in many ways. All probably have small, rocky cores, and all are shrouded by thick clouds so that all we can observe directly are the outer layers of their atmospheres. All are surrounded by rings, but only Saturn has an extensive system of bright rings that can be observed from Earth. Jupiter and Saturn have at least sixty moons each. Uranus has twenty-seven moons and Neptune has fourteen. All but the largest and brightest of the moons were discovered by spacecraft during flybys.

Exotic Moons

In many ways, the moons are even more interesting and diverse than the planets. There are erupting volcanoes and geysers, methane lakes, and possibly even places where life might have developed. ●

▲ **Jupiter's four Galilean moons.** From top to bottom they are Io, Europa, Ganymede, and Callisto. This is also the order of their distance from Jupiter, with Io being the closest. Ganymede is even larger than Mercury and is the largest moon in the solar system. Recent observations suggest that it may have a saltwater ocean hidden beneath its crust. Callisto is the most heavily cratered object in the solar system. These craters were mostly formed about 4 billion years ago, and there have been no subsequent lava flows to cover them up. Thus, we are seeing the surface of Callisto as it was 4 billion years ago, unaltered by subsequent erosion or lava flows. NASA Planetary Photojournal

▲ **The surface of Europa.** The colors are approximately what the human eye would see. Long, linear cracks and ridges criss-cross the icy surface. Areas that appear blue or white contain relatively pure water ice, while reddish and brownish areas include non-ice components. Evidence suggests that there is a liquid ocean beneath the icy surface crust. NASA is developing concepts for a spacecraft that would make repeated flybys past Europa in order to measure its surface composition and, by means of radar, determine the thickness of the icy crust. This may be the best place in the solar system to look for present-day life beyond Earth. NASA/JPL-Caltech/SETI Institute

◀ **An eruption of the volcano Tvashtar Catena on Jupiter's moon, Io.** In this false-color image, white and orange areas on the left side of the picture show newly erupted hot lava. The bright orange material is cooler than the material shown in white. The two small bright spots are sites where molten rock is exposed at the toes of the lava flows. The larger orange and yellow ribbon is a cooling lava flow that is more than 37 miles long. The dark L-shaped lava flow was erupted a few months earlier and has now cooled off. This image is about 155 miles across. Many astronomers can still remember how surprised they were when they saw the first image of an eruption on Io. This was the first evidence that any bodies in the solar system other than the Earth still had volcanic activity. NASA Planetary Photojournal

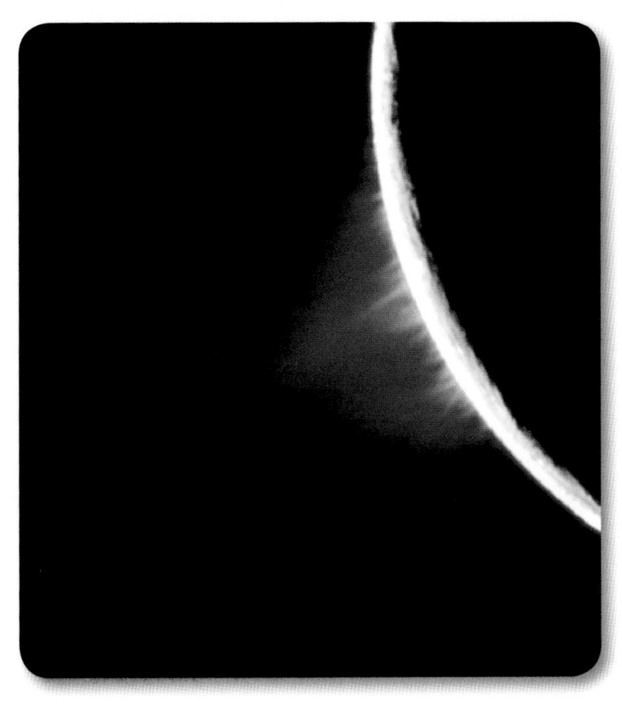

▲ **Geysers on Saturn's moon, Enceladus.** The blue color was chosen for dramatic effect. In addition to water, these geysers contain other volatiles, including hydrogen, carbon dioxide, ammonia, and methane. Scientists now believe that Enceladus harbors an internal liquid-water ocean under tons of icy crust, and that the plumes originate there. The eruptions appear to be continuous, and the particles and gas are rushing outward at a speed of 800 miles per hour. NASA/JPL/Space Science Institute

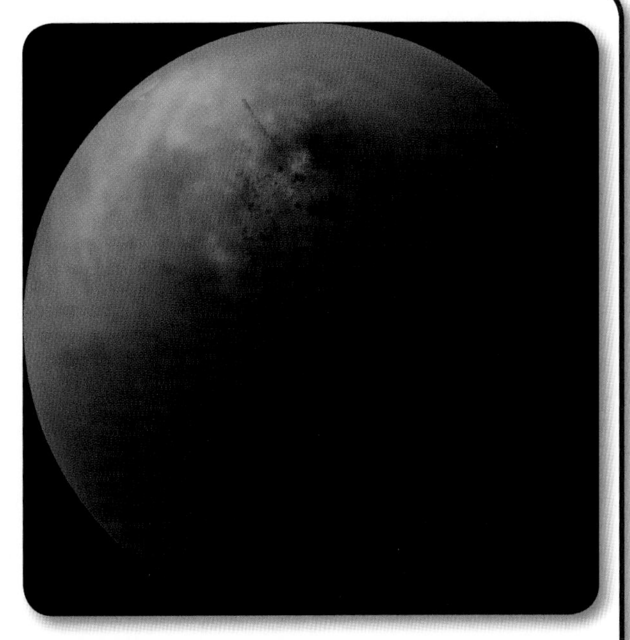

▲ **False color mosaic image, showing lakes on Saturn's moon Titan.** Kraken Mare, Titan's largest sea, is about the size of Lake Superior and the Caspian Sea combined. The orange areas are thought to be places where the liquid has evaporated—Titan's version of the salt flats in the deserts of Earth's Southwest. Titan's "rocks" are made of water ice. These images were made with infrared cameras, which can see through Titan's thick clouds. NASA/JPL-Caltech/University of Arizona/University of Idaho

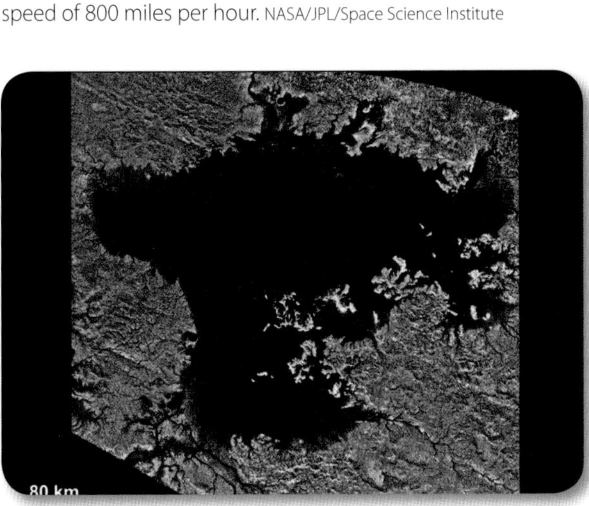

◀ **Ligeia Mare, the second-largest known body of liquid on Saturn's moon, Titan.** It is filled with liquid hydrocarbons, such as ethane and methane, and is one of the many seas and lakes that bejewel Titan's north polar region. In this false-color image, liquids appear black, and the solid surface of Titan appears yellow. NASA/JPL/Space Science Institute; NASA/JPL-Caltech/ASI/Cornell

EXPLORING ON YOUR OWN

The Planets

The planets Venus, Mars, Jupiter, and Saturn are some of the brightest objects in the night sky. Many sources will tell you which ones, if any, are visible on any given night. Try your local paper or the websites for *Astronomy* or *Sky and Telescope* magazines. Free apps for a smartphone or tablet will tell you whether any specific planet is above or below the horizon and for those above the horizon will tell you exactly where to look. Mercury is bright but elusive because it is always low on the horizon and can be seen only immediately after sunset or before sunrise. An app will be particularly helpful if you want to be one of the few people who have seen Mercury.

The best way to explore what we know about the planets is to take advantage of the rich resources provided by NASA. Start by going to https://solarsystem.nasa.gov/planets/, which has information about each planet, as well as about dwarf planets and comets. There are extensive image galleries, activities and games for kids, and lesson plans for teachers. ●

Dwarf Planets: Why Pluto Isn't a Planet Anymore

Pluto was discovered in 1930 by Clyde Tombaugh at Lowell Observatory. For six decades after that, except for Pluto, the outer solar system beyond Neptune was thought to be essentially empty. It wasn't until the 1980s that a planetary astronomer, Dave Jewitt, began to wonder whether this apparent emptiness was real or whether astronomers simply had not had the right equipment to do an effective search. Any objects out there were likely to be no more than a few hundred miles in diameter, and at such a large distance, they would also be extremely faint. In 1986, Jewitt and Jane Luu began a systematic search for objects beyond the orbit of Pluto. The search produced no discoveries for six years. Then in 1992, taking advantage of the excellent seeing on Mauna Kea and improved detectors, they had their first success.

Scientists now estimate that there are at least 70,000 objects beyond the orbit of Neptune with diameters larger than 60 miles. The region occupied by these objects is called the Kuiper Belt. Observations of a few of the brightest objects suggest that they are composed of frozen carbon monoxide (CO), methane (CH_4), and other ices mixed with dirt.

Not all of the Kuiper Belt objects are small. One, now officially named Eris, has almost the same diameter as Pluto, and there are several others that are not much smaller. So now astronomers had only two choices: would there be more than nine planets? Or should "planet" be redefined in such a way that Pluto was no longer a planet? In 2006, the International Astronomical Union decided that a planet had to have the following characteristics:

These artist's concepts show some of the best-known objects in the Kuiper Belt in comparison to Earth. The image was labeled before the objects got their official names. With their official names, the objects in the top row, left to right are: Eris and its moon, Dysnomia; Pluto and its moon, Charon; and Makemake. In the bottom row, left to right are Haumea with its moons, and Sedna and Quaoar. Data from the New Horizons flyby of Pluto show that its diameter is actually about 25 miles larger than the diameter of Eris, but Eris is about 27 percent more massive than Pluto. This means that Pluto has a higher proportion of ice to rock. A recent analysis of new sky survey data by Mike Brown, who with collaborators at Caltech has discovered most of the dwarf planets, suggests that it is unlikely that any more objects as bright as these are still to be found. NASA

1. It must orbit the Sun.
2. It is not a Moon.
3. It is massive enough to be round.
4. It must have cleared its orbit.

The first two are easy to understand. The requirement that the object must be round rules out really small rocky bodies like asteroids. If a body has enough mass, then gravitational forces will build up pressure in its interior so that the material becomes plastic enough to be deformed to take on a round shape. Anything larger than 300 miles in diameter is highly likely to be round.

Pluto fails only the fourth test. It orbits in a region where there are many other objects of similar size. The orbital paths of the eight planets in the solar system are empty of other large objects. That is because a true planet, according to this definition, exerts enough gravitational force either to attract and accrete nearby debris or to kick small objects out of its path and send them to large distances far away from its orbit.

While this is a precise, clear, scientific definition, many astronomers think that Pluto should have been grandfathered in and permitted to officially remain the ninth planet.

Will actually seeing Pluto change any minds? Now, nearly a decade after its "demotion," we have finally seen detailed images of Pluto for the first time. It turns out to be a complicated world with many of the properties found on the other planets in our solar system: five known moons; complicated geology, including mountains that are young by geological standards; an atmosphere and seasons; a surface covered by rock and ice; and flowing glaciers.

In July 2015, the New Horizons spacecraft zipped past Pluto at a speed of more than 30,000 miles per hour and gave us our first close-up view of this distant world at the edge of our solar system. At its closest approach, the spacecraft was only 7,750 miles above the surface of Pluto. To reach Pluto, the New Horizons spacecraft traveled 3 billion miles, a journey that required nearly 9.5 years. As this is written, 10 days after the flyby, only about 5 percent of the data collected have been returned to Earth. The rest will trickle in over the next 12 to 18 months.

Why the agonizing wait? Sunlight at the distance of Pluto is too weak to provide power through solar cells. Instead, the New Horizons spacecraft is powered by the heat from the natural radioactive decay of plutonium dioxide. This heat is used to generate electricity. When the flyby occurred, the power available to the spacecraft was equivalent to that produced by two 100-watt light bulbs. Because of the limited power available, it was not possible to take data and simultaneously transmit it back to Earth. All of the observations were preprogrammed, and the data were then stored aboard the spacecraft. Scientists and the public had to wait several hours to see if any observations were successful.

A false-color image of Pluto's surface. True colors are exaggerated to show differences in composition. The smooth, light-colored heart-shaped region, named after the discoverer of Pluto, is referred to as Tombaugh Regio. It is covered with a combination of methane, nitrogen, and carbon monoxide ice. The regions of similar color that extend down toward the bottom of the image from the heart are probably produced by ice flowing out of the heart-shaped region.

The dark, reddish patches in this picture are probably complex molecules formed from simple hydrocarbons. Pluto has a hazy atmosphere that contains methane gas particles, which can be broken up by ultraviolet sunlight, even at the distance of Pluto. The newly freed components of methane—hydrogen and carbon—can then recombine to form molecules like ethylene and acetylene. As these hydrocarbons fall to lower, colder parts of the atmosphere, they condense into ice particles. When irradiated by sunlight, these simple hydrocarbons can form complex molecules that appear dark and reddish. NASA/JHUAPL/SwRI

Because of limited power, the radio signals sent back to Earth are very weak—so weak that New Horizons can send back only 126 bytes per second—compare that with the megabytes in an image taken with a smart phone! Even with the use of lossless compression, it takes 42 minutes to send a single 1024 × 1024 image back to Earth. Only three radio antennas in NASA's Deep Space Network are sensitive enough to detect these weak signals.

With 95 percent of the data still to be returned, most of the scientific analysis remains to be completed, but the first few images tell a fascinating story. We can already see that Pluto has icy mountain ranges that reach altitudes as high as 11,000 feet above the surface. It also has flat plains covered with methane, nitrogen, and carbon monoxide ices. Its surface temperature is about -390 degrees Fahrenheit.

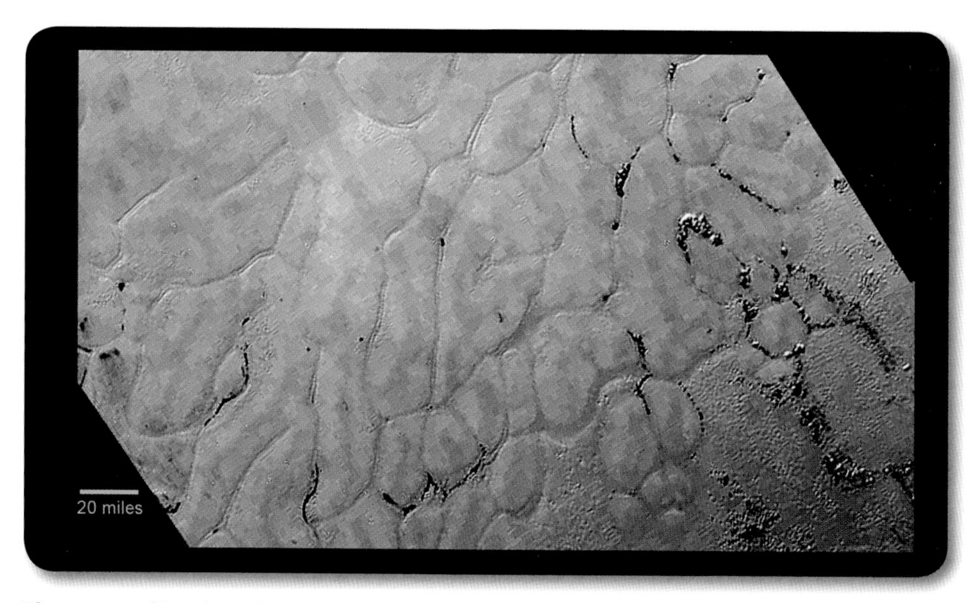

The center of Tombaugh Regio. The absence of craters suggests that the ice has been deposited here no more than 100 million years ago. This plain has been named Sputnik Planum, after Earth's first artificial satellite. The surface appears to be divided into irregularly shaped areas, which are separated by troughs. These shapes may be produced as material on the surface contracts—for an analogy, think of dried mud flats in the deserts of the Southwest. Alternatively, tiny warmth from the interior of Pluto may cause convection of the surface ices—think of the blobs of wax that rise slowly in a lava lamp. NASA/JHUAPL/SwRI

Scientists can estimate the ages of planetary surfaces by counting craters. Old surfaces have been bombarded for billions of years by asteroids and have large numbers of craters. Our Moon and Mercury are examples. Surfaces that have been recently covered by lava or ice have few craters. Some parts of Pluto's surface have so few craters that they must be young, geologically speaking. Some of the mountains have apparently been uplifted no more than 100 million years ago, and the smooth, icy plains may be only a few tens of millions of years old. Pluto itself is about 4.5 billion years old.

Flows of new mountain-building ices from the interior of the planet to the surface must be driven by heat, but what is the source of heat on frigid Pluto? Scientists have suggested two possibilities: the release of heat from radioactive elements, as occurs in the interior of the Earth; or the release of heat when liquids freeze to form ice. Does Pluto have a liquid ocean buried deep in its interior? That is just one of many questions that can be answered only after much more detailed scientific analysis.

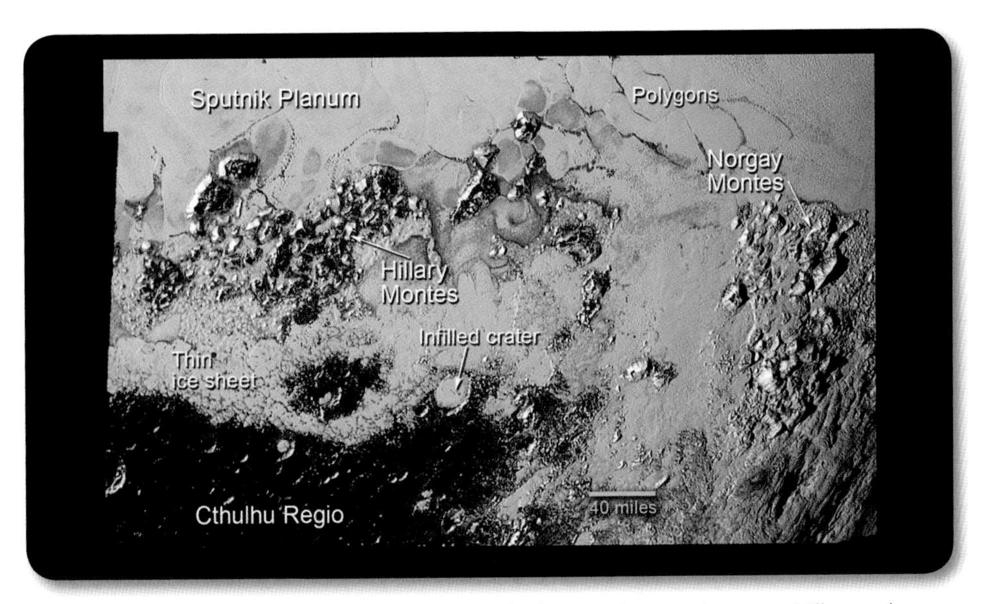

The southern edge of Sputnik Planum. The newly christened mountain ranges, Hillary and Norgay Montes, are named after the two men who first summited Mount Everest. These mountains on Pluto are, however, much more modest in altitude. Only about a mile high, they are similar in height to the Appalachian Mountains. Some parts of the surface are heavily cratered, an indication that they are billions of years old. The bright, smooth areas are covered with ice; the absence of craters indicates that the ice has been deposited on the surface recently, geologically speaking—probably in the last few tens of millions of years. This image shows that ice has apparently flowed into the cratered terrain. The crater that is filled with ice is about 30 miles wide. NASA/JHUAPL/SwRI

There is so much more to learn. Keep track of the latest discoveries at nasa.gov/newhorizons.

It Came from Outer Space

It Came from Outer Space is the title of a 3-D black and white movie made in 1953. It tells the story of an astronomer and his fiancée who are stargazing in the desert when they see the fiery crash of an alien spacecraft.

That's just fiction, you say. Well, it did happen here—without the aliens, of course. There were no humans here to see it, but about 50,000 years ago a meteorite crashed into Northern Arizona. Meteor Crater east of Flagstaff is the place where this occurred. The meteorite was made of nickel-iron and was about 50 meters (about 165 feet) in diameter.

Aerial view of Meteor Crater in northern Arizona. This is one of the youngest and most accessible craters on Earth. Shortly after its formation 4.5 billion years ago, the Earth was bombarded by meteorites. Those ancient craters have been eroded or covered over by lava flows or destroyed by the geologic processes associated with plate tectonics. Meteor Crater is almost a mile across and more than 550 feet deep. U.S. Geological Survey

It impacted the Earth at a speed of about 20 miles per second and released about 10 megatons (million tons) of energy. For comparison, the bomb over Hiroshima was almost a thousand times less powerful (15 kilotons or 15 thousand tons of energy).

The extinction of the dinosaurs 65 million years ago was probably caused by an even larger impact, with perhaps some contribution by massive volcanic activity in India. This Cretaceous-Tertiary (K-T for short, with the K from the German word for Cretaceous)

extinction not only eliminated the dinosaurs that previously roamed the Earth but also caused mass extinction of about 70 percent of all the plant and animal species then living on the Earth. The trigger for the extinction was the impact of an asteroid about 30 miles in diameter; remnants of the resulting crater, which is about 112 miles in diameter, have been found off the coast of Yucatan. The impact of such a large asteroid would have caused a global firestorm, earthquakes, and tsunamis. Huge quantities of dust thrown up into the atmosphere would have dimmed sunlight and caused a global winter that might have lasted for years.

Perhaps the biggest impact of all occurred about 150 million years after the Earth formed. A small planet about the size of Mars struck the Earth, blasting a huge quantity of rock into space. This cloud of rocks eventually coalesced due to the force of gravity and became Earth's Moon.

But that is all ancient history, you say. Surely, it can't happen again. In fact, as recent events have shown, the Earth's neighborhood remains a dangerous place. Some examples:

In 1908 over Tunguska, Siberia, a large meteoroid or comet hit the atmosphere at high speed and exploded in the air about 3–6 miles above the Earth's surface with an energy of 10–15 megatons. This air burst flattened about 80 million trees over 830 square miles.

On October 7, 2008, an asteroid exploded about 23 miles in the air above the Nubian Desert in Sudan. Scientists estimate that the asteroid was about 13 feet in diameter and weighed almost 90 tons. This asteroid was actually discovered 19 hours before impact by Richard Kowalski at the Catalina Sky Survey, which uses a telescope just north of Tucson. This was the first time that an incoming asteroid was detected prior to impact.

An even more spectacular event occurred on February 15, 2013, near Chelyabinsk, Russia. First observed as a large fireball, this incoming asteroid was a relatively small one—only 60 feet in size with a weight of over 11,000 tons—that entered the atmosphere at over 40,000 miles per hour. The asteroid shattered at high altitude, and pieces of various sizes reached the ground as meteorites. Damage to buildings was estimated at 33 million dollars and about 1,500 people sought treatment for injuries, mainly from broken glass.

We have labeled all of these impactors as *asteroids*, but just what is an asteroid? They are objects that orbit the Sun, just as the planets do. Some are small, others are over a mile in size. (Even larger asteroids are found in the asteroid belt between Mars and Jupiter.) Some contain metals, like iron and nickel, but most are made of rocky material like basalt and olivine and possibly some ice. They can be solid or piles of rubble. (Comets also contain some dust and rocky material but much larger quantities of ice that can be vaporized when heated by the Sun. It is this vaporized material that forms the tail of a comet.)

Eros, an asteroid with an orbital period around the Sun of 1.76 years. Its dimensions are 21.4 × 7.0 × 7.0 miles, and the craters on its surface show that it is solid rock, not a pile of rubble. In 2001, NASA landed a space probe on its surface after taking many images from orbit. In 2012, Eros came within 16.6 million miles of the Earth—no danger since it was about 80 times farther away than the Moon. NASA/JHUAPL

What Can We Do to Mitigate the Risk of Impact?

Motivated by these recent impacts, the U.S. Congress has mandated that NASA determine if there are other potentially hazardous asteroids (PHAs) that might do significant damage if they were to strike the Earth. As a result, surveys to discover near-Earth asteroids have been undertaken at several observatories. The ones currently most active are the University of Hawaii's PanSTARRS survey and the University of Arizona's Catalina

Sky Survey. The WISE Observatory also conducted a brief survey from space. As a result mainly of the ground-based searches, the rate of discovery of near-Earth asteroids has risen from fewer than twenty annually in 1995 to more than six hundred each year. The Large Synoptic Survey Telescope, with its large aperture and large field of view, will be a very efficient machine for discovering asteroids.

Not all asteroids that pass near the Earth are hazardous. To determine whether a particular asteroid is likely to be dangerous, it is necessary to make several observations of it as it moves through the sky to discover whether its orbit intersects that of the Earth.

Suppose we do find an asteroid that is on a collision course with Earth. What can we do about it? There are basically two options: a) civil defense strategies to manage the consequences, or b) send a spacecraft out to change the orbit. The choice depends on the size of the object and the amount of warning time. For the smallest and largest objects or a threat from any object with very little warning, all that can be done is to estimate where the impact will take place and then evacuate people.

If a dangerous object is discovered decades before impact, there are several ideas being explored to change the orbit. The "slow push" method would work for small asteroids. A spacecraft would be parked near the asteroid and its gravitational attraction would change the orbit. One or more spacecraft crashing into asteroids that are up to a kilometer in size could change their orbits enough to avoid collision with Earth. Nuclear explosion might be the only way to deflect truly large asteroids. In this case, the explosion would be set off close to the asteroid. The explosion would vaporize the surface, and the main body would recoil, thus changing its orbit.

All of this sounds like science fiction. How worried should you actually be? Small objects are by far the most common and do the least damage. Really large impacts are exceedingly rare. What are the chances that there is a really large asteroid headed toward Earth that we haven't yet detected? The total number of near-Earth asteroids was recently estimated from observations with NASA's Wide-field Infrared Survey Explorer (WISE). There are likely to be fewer than 1,000 asteroids with diameters large enough to affect the Earth's climate. More than 90 percent of these have already been found, and none is headed for Earth. Impacts like the one over Chelyabinsk may be expected on average once every 150 years or so. So even if we haven't yet found all of the large asteroids, don't lose too much sleep over this risk—large, dangerous impacts are rare.

Space Weather

There is an even more immediate risk from outer space—space weather. The term *space weather* may sound like a contradiction. Space is, after all, a vacuum, so how can it have weather? Space weather refers to changing conditions in space, mainly as a result of giant

Mission to an Asteroid

Each NASA mission has a dedicated website. OSIRIS-REx is the first U.S. mission that will obtain a sample from an asteroid and return it to Earth. You can track the progress of OSIRIS-REx by going to www.asteroidmission.org or by simply typing the name of the mission into a search engine. The site describes the scientific goals, the timeline for the mission, why Bennu was the specific asteroid chosen for the mission, etc. The entire mission from planning and fabrication of the spacecraft to analysis of the sample will take 14 years (2012–2025), which illustrates just how much patience a space scientist must have! ●

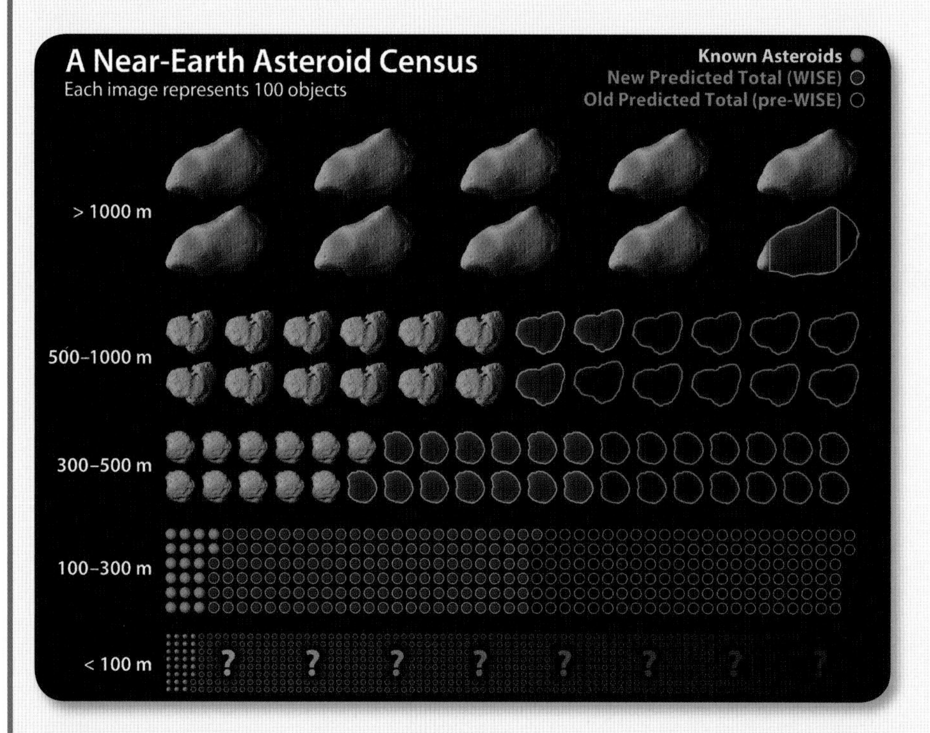

▲ **This chart summarizes what we know about the number of near-Earth asteroids.** Each asteroid in the image represents about 100 actual asteroids. Near-Earth asteroids already discovered are shown in brown. The green outlines show an estimate based on WISE data of how many asteroids in each size range are so far undiscovered. The blue outlines show an older estimate of the number of undiscovered asteroids.

This chart shows: 1. There are estimated to be slightly fewer than 1,000 asteroids that are larger than 1 kilometer (0.6 miles) in diameter; of these, 911 have already been discovered and present no danger; 2. Smaller asteroids are much more numerous than large ones; and 3. The new data indicate that there are fewer near-Earth asteroids than previously thought. This study does not apply to objects smaller than 100 meters (330 feet), but it is estimated that there are more than a million in this size range based on previous studies. NASA/JPL-Caltech

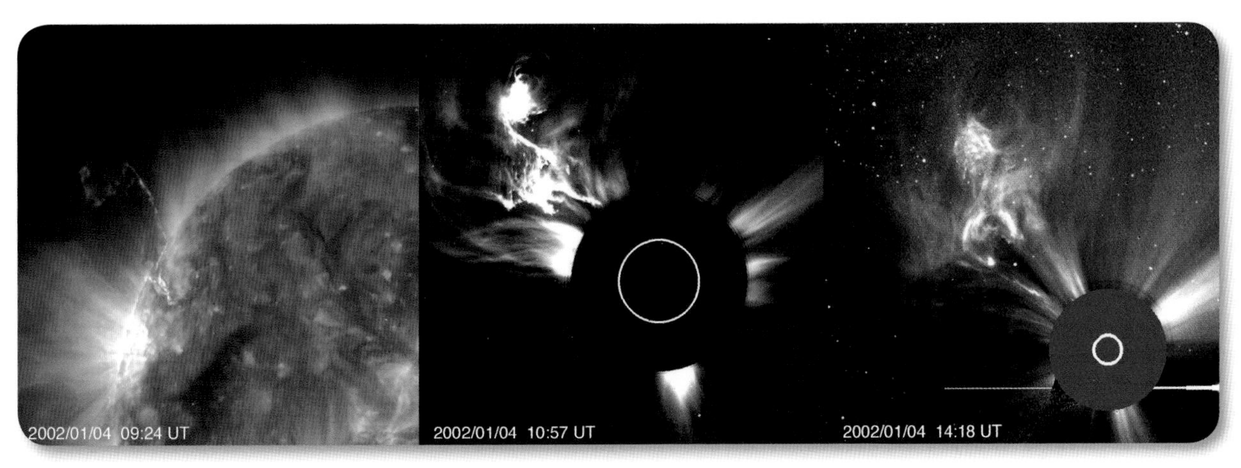

A spectacular coronal mass ejection (CME) as viewed by three different space instruments. Note the change in scale in the three images. The left image shows the rim of the Sun. The white circles in the other two images show the diameter of the Sun, which is 860,000 miles. The dark disks have blocked out the glaring light of the Sun's disk so that we can see the fainter exploding gas streams. The times given in each image show that this gas explodes away from the Sun in only a few hours. In this case, the gas was ejected at almost 90 degrees away from Earth and did not affect the Earth's space weather. SOHO/EIT and SOHO/LASCO (ESA & NASA)

storms on the Sun. Modern society, with its dependence on sophisticated electronics, is increasingly vulnerable to damaging effects from solar storms.

The largest of the solar storms are the coronal mass ejections (CMEs). These storms eject millions of tons of gas, mainly protons and electrons, at speeds up to about a million miles per hour. On average, these ejected particles reach the Earth's orbit in about 3.5 days. Some of the particles penetrate the Earth's atmosphere where they interact with oxygen and nitrogen atoms and cause auroras (Northern and Southern Lights) in regions around the Earth's magnetic poles. The Sun itself has a cycle of activity that lasts on average about 11 years, and auroras are more common when solar activity is at its peak. The best places to observe auroras are places well to the north—Alaska, Iceland, and northern Scandinavia. Rarely, exceptionally strong auroras have been seen as far south as Arizona.

The Earth's magnetic field extends from the center of the Earth out into space. The particles ejected by CMEs interact with this magnetic field and cause changes to it. Changing magnetic fields cause electrical currents to flow, and some CMEs have induced currents high enough to melt the copper windings in transformers, which are essential components of power distribution systems. A solar storm in 1989 caused 6 million residents of Quebec to lose power for nine hours.

Electronics aboard spacecraft are vulnerable to damage by CMEs. Space weather events can cause the ionosphere surrounding the Earth to expand, thereby increasing frictional drag on Earth-orbiting satellites, changing their orbits. Changes in the ionosphere can also distort signals from GPS satellites. These effects can be large enough to reduce the accuracy of GPS positions and even to disable the precision navigation tools used by the Federal Aviation Administration (FAA) for commercial aviation. This has led to flight restrictions for minutes or in a few cases even for days because the FAA requires airplane systems to have GPS positions accurate to 50 meters (about 160 feet).

Astronauts in near-Earth orbit are partly shielded from harmful radiation by the Earth's magnetic field, but astronauts voyaging to Mars will not have that protection. Because of weight restrictions, spacecraft are not built with radiation shielding. Instruments on unmanned Mars missions are already making measurements that will be used to estimate whether the amount of exposure that would be expected on a manned mission is acceptable.

There are now many government organizations and companies developing ways to mitigate the effects of solar storms. Infrastructure can be designed to be more resistant to the currents generated by disturbances in the Earth's magnetic field. There are also efforts to improve forecasting of space weather. With accurate predictions, utilities and operators of systems like satellites with sensitive electronics can take precautions by disconnecting systems and turning off critical hardware.

An aurora viewed from Kitt Peak on March 28, 2001. This was a year when solar activity was at a maximum. This auroral display lasted about an hour. Adam Block/NOAO/AURA/NSF

EXPLORING ON YOUR OWN

Space Weather

Type "space weather" into a search engine, and you will find a number of sites devoted to this subject. Some are quite technical and intended for commercial users. Other sites interpret space weather more broadly to include anything interesting happening in Earth's space environment. On these sites, you can find, among other things, real-time images of the Sun, information about its level of activity, and the current status of the solar cycle. You can also find data about near-Earth asteroids currently passing close by, information and images about recent lunar and solar eclipses, real-time images of auroras from around the world, and so forth. The combination of "Iceland" or "Norway" with "aurora" will yield especially stunning images.

NASA's Jet Propulsion Laboratory (JPL) maintains a site that offers an excellent introduction to research about the discovery and analysis of near-Earth objects (http://neo.jpl.nasa.gov). ●

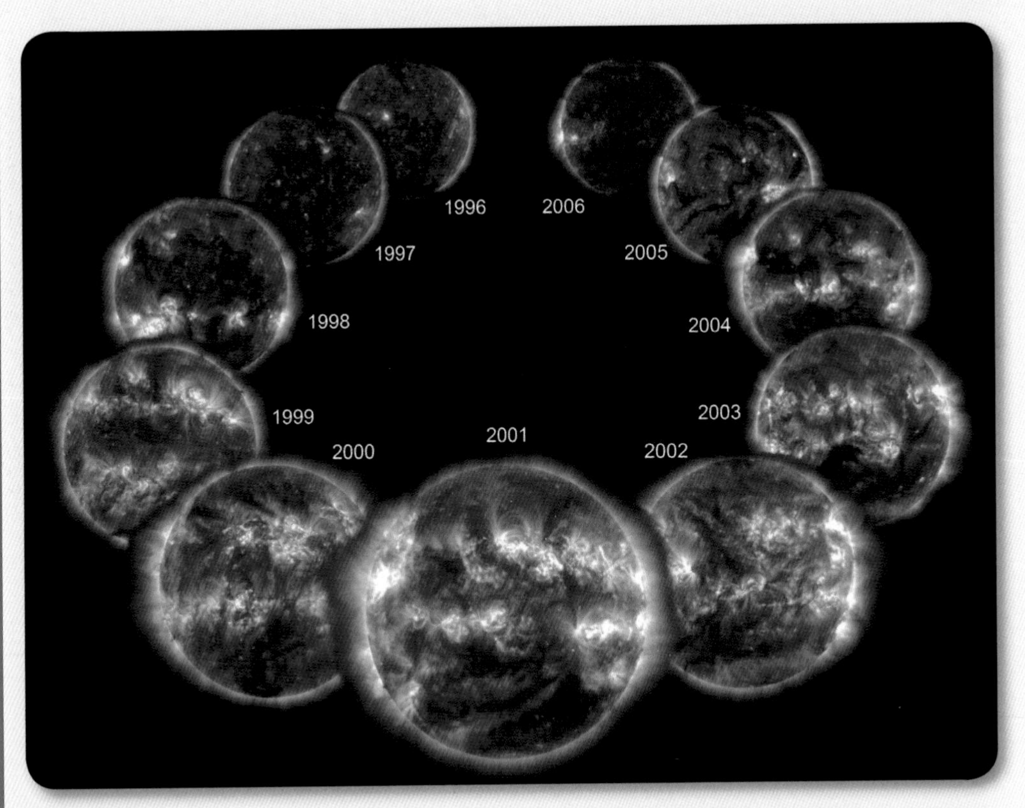

▲ **An image taken in ultraviolet light from each year of nearly an entire solar cycle.** The bright parts of these images correspond to gas in the Sun's atmosphere at a temperature of about 1.5 million degrees Kelvin. The amount of hot gas, the number of sunspots, and other indicators of solar activity vary on a cycle that lasts on average 11 years. Note that the solar activity was near maximum when the aurora was seen far south over Kitt Peak in Arizona. NASA GSFC

▲ **An image of the aurora australis or southern lights.** Auroras occur when the solar wind causes charged particles, mainly protons and electrons, to plunge into the atmosphere of the Earth. These charged particles then collide with, and add energy to, atoms in the upper atmosphere. Green auroras are produced when oxygen atoms at altitudes above about 50 miles give up this added energy. This image was taken by astronauts as the International Space Station (ISS) passed over the Indian Ocean between Madagascar and Australia. Solar panels and other parts of the ISS can be seen in the upper right of the image. NASA's Earth Observatory

The Earth and a solar eclipse of the Sun imaged by the Cassini spacecraft as it orbited Saturn. Earth appears as a tiny pale blue dot at about the ten o'clock position on the left-hand side of the image just above Saturn's bright rings. We have images of only a few planets outside the solar system, and like the Earth in this image, all appear as mere points of light. In this image, the night side of Saturn is partially illuminated by light reflected from its majestic ring system. The rings look dark when silhouetted against Saturn. Away from Saturn, the rings look bright because they scatter sunlight. Cassini Imaging Team, SSI, JPL, ESA, NASA

5 The Search for Planets Around Other Stars

One of the hottest topics in astronomy these days is the search for life elsewhere in the Universe. Does life exist anywhere except Earth? The answer to that question depends on a number of factors:

- Are there planets around other stars?
- If planets do exist, how many of them have reasonable enough temperatures to allow life to develop?
- And if life does develop, is there a way to detect its presence from here on Earth?

Recent research has made great strides in answering the first two of these questions. The discovery of evidence of life itself is likely many years in the future, but we will look at some of the strategies that might work.

It wasn't until 1995 that we discovered the first planet orbiting a star like our Sun. The numbers of such planets have skyrocketed since then, mainly as a result of observations by the Kepler satellite. Let's look first at the techniques used to find exoplanets, which is the term for planets beyond the solar system.

Seeing is believing, but it is actually very difficult to obtain an image of an exoplanet. A typical planet is a million or so times fainter than the star around which it orbits. Trying to see such a planet is somewhat like trying to detect a mosquito in the glare of a distant streetlight. As a result, we have images of only a handful of planets.

More than 1,500 planets that we haven't actually seen have been discovered by measurements of the effects of the planets on their parent stars. A planet can affect its parent star in two ways. It can cause the velocity of its sun to change with time. And if we view the planet from just the right direction, it can also block some of the light from its sun and make it look dimmer.

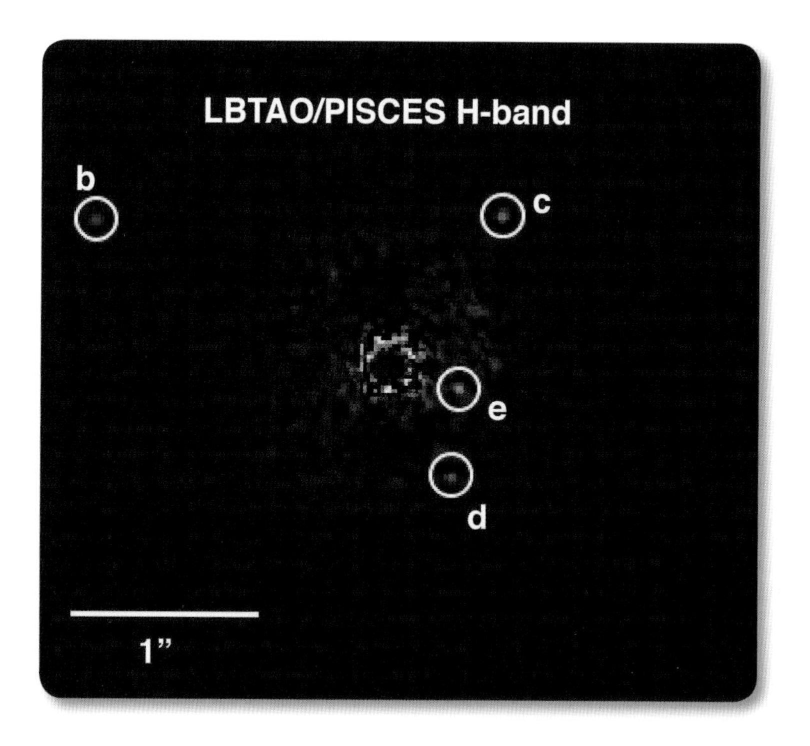

An infrared image of exoplanets taken with the adaptive optics system on the Large Binocular Telescope. These four planets (b, c, d, and e) orbit the star HR 8799. The light from the star has been blocked out by the instrument so that the planets, which are much, much fainter than the star, can be seen. With current instrumentation, we can obtain images of only a few planets, and in even the best images, the planets will look like mere points of light.
LBT Observatory

Discovery of Exoplanets: The Doppler Effect

To understand how a planet can affect the motion of a star, suppose we have a system with a central star and one planet. Usually, we say the planet orbits the star, but in fact it is also true that the star orbits the planet. The more correct statement is that both the planet and the star orbit around a point called the center of gravity. Imagine the star and the planet are at either end of a seesaw. The point where a support would exactly balance the seesaw is called the center of gravity. Obviously that point has to be much, much closer to the star, which is much, much more massive than the planet. Think of trying to balance a 350-pound man on a seesaw with a 5-year-old child on the other end! In the case of the solar system, the Earth by itself causes the Sun's velocity to vary by only 10 centimeters per second or about 0.23 miles per hour. Jupiter, the most massive planet in our solar system, causes the Sun's velocity to vary by about 29 miles per hour.

We use spectroscopy to measure the velocity of a star. If we take the light from a star and spread it out into its rainbow of colors, we find that there are a large number of dark lines where light is missing. The dark lines are produced when gases in the star's atmosphere absorb light from hotter regions deeper inside the star. Just as each human being has a unique set of fingerprints, each type of atom—hydrogen, calcium,

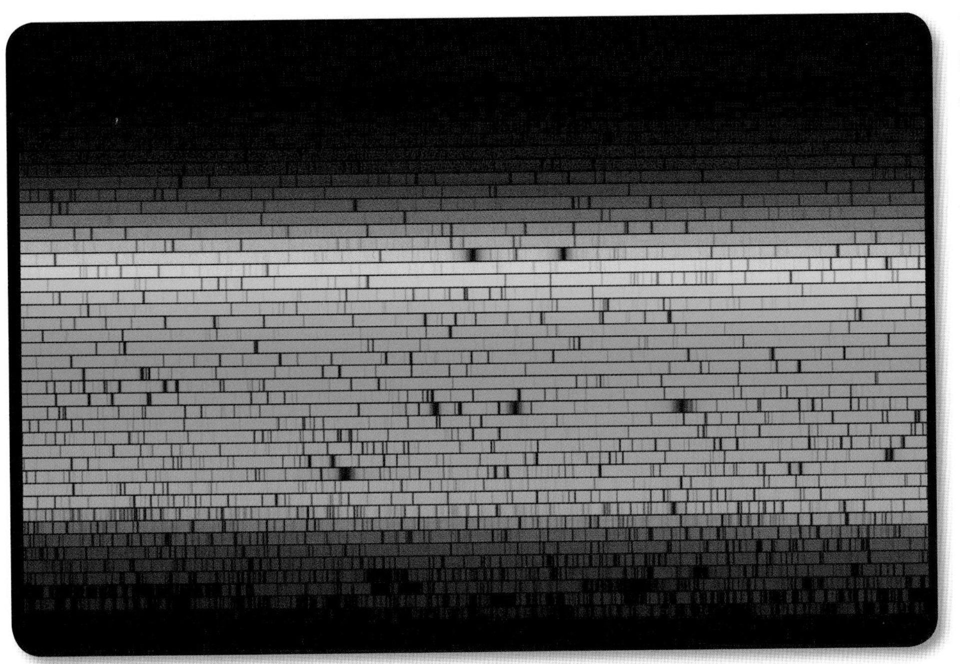

The spectrum of our Sun. This image was created by sending the visible light of the Sun through an instrument that spread that light out into the rainbow of colors that together combine to produce the white light of the Sun. Wavelengths increase from blue to red and from left to right along each strip. The dark lines crossing the spectrum show where light has been absorbed by elements in the Sun's atmosphere. Each element has a specific pattern of lines—its own unique fingerprint. For example, the strongest dark line in the deep red portion of the spectrum is produced by hydrogen. The two strong lines in the yellow about one-third of the way down from the top are produced by sodium. The most abundant element in the Sun is hydrogen, and helium is second. Most of the elements found on Earth, including carbon, nitrogen, oxygen, and iron, have also been detected in the Sun's atmosphere.
N. A. Sharp, NOAO/NSO/Kitt Peak FTS/AURA/NSF

carbon, iron, etc.—has a unique pattern of lines. That is how we tell what stars are made of.

Remember that light is a wave. If an object that emits waves is moving toward you, the wave crests will be more closely spaced than if an object is stationary. If that object is moving away from you, the wave crests will be spaced farther apart. Sound is a wave, and you have probably heard this effect. Listen to a fire engine siren as it approaches you and note its pitch. You will hear the pitch drop when the fire engine passes you and begins to move away. This happens because the sound waves are more closely spaced and have a shorter wavelength when the fire engine is approaching you. Shorter wavelength corresponds to higher pitch. This effect was first demonstrated in 1842 by Christian Doppler, who hired musicians to play on an open railroad car as it moved along a track.

The Doppler effect applies to light as well. If a star, for example, is moving toward us, the lines in its spectrum will be shifted to shorter, bluer wavelengths. If it is moving away from us, the lines will be shifted to longer, redder wavelengths. Measurements of these Doppler shifts provided the first evidence of exoplanets.

Massive planets close to their parent stars produce the biggest changes in the star's velocity, and it is fairly easy with current techniques to detect such planets. Astronomers, however, continue to improve their instrumentation with the goal of being able to detect

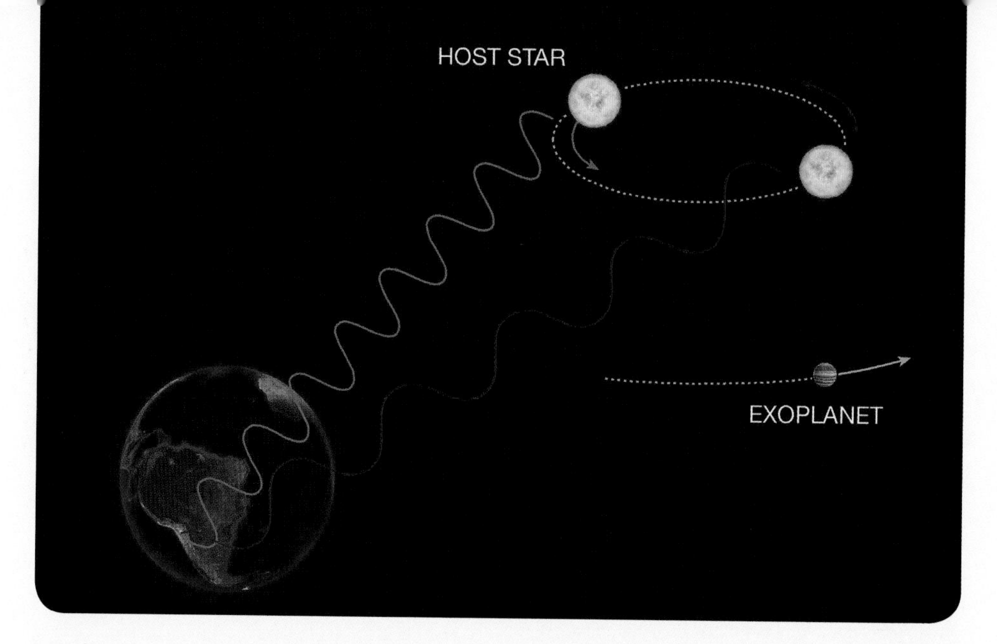

The motion of an exoplanet's parent star as seen from the Earth. The gravitational force exerted by a planet causes its sun to move in a small orbit around their common center of gravity. As the star moves around its orbit, the lines in its spectrum will be shifted alternately toward the blue (shorter wavelengths) when the star is moving toward us, and toward the red (longer wavelengths) when the star is moving away from us. By regularly looking at the spectrum of a star, astronomers can measure its velocity and see if it moves due to the influence of a planet. These changes in wavelength are very small and very difficult to measure. ESO Press Photo 22e/07 (25 April 2007)

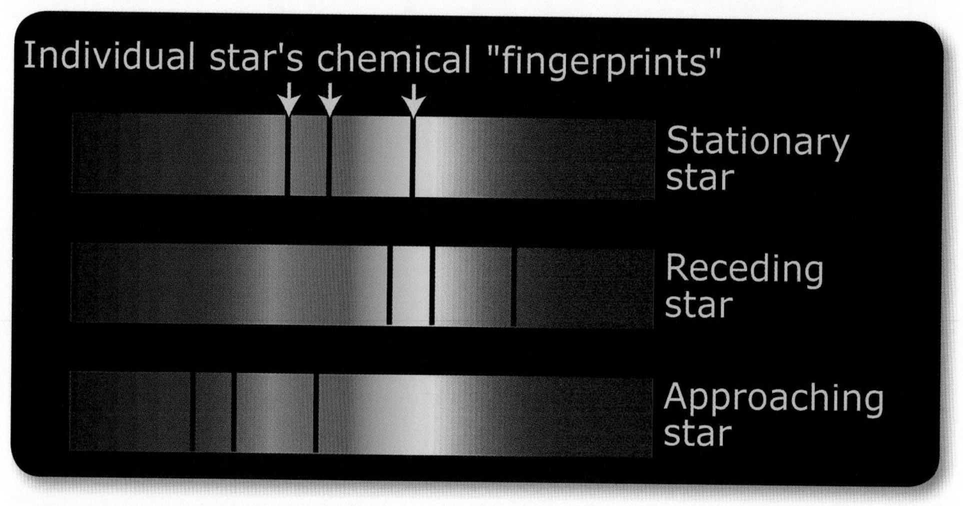

Spectra showing the Doppler effect. The top spectrum shows where the dark lines of an imaginary element would be located in the laboratory or in a star if it were not moving. The middle spectrum shows that the lines would be shifted toward the red if the star were moving away from us. The lines would be shifted toward the blue if the star were moving toward us. The amount of the shift caused by an exoplanet is very much smaller than the shifts shown here. STScI

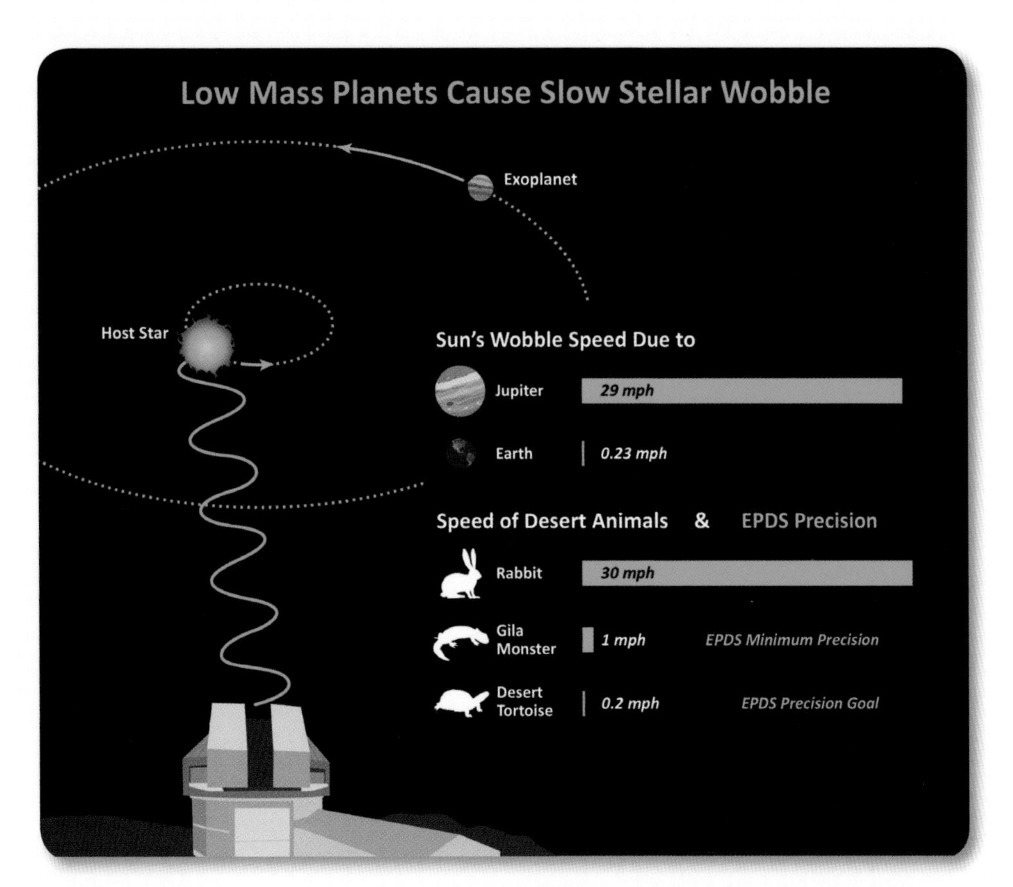

Low-mass planets cause small changes in the observed velocities of the stars around which they orbit. NASA is funding the construction of an Extreme Precision Radial Velocity Spectrometer that will be installed on the WIYN Telescope on Kitt Peak. This instrument will be able to detect the effects produced by planets with masses similar to those of Jupiter and Neptune. It may even be able to discover Earth-sized rocky planets. This image shows that Jupiter causes the Sun to move around their common center of gravity at a velocity of 29 mph. In contrast, if the Earth were the only planet in our solar system, it would cause the Sun to move— and its spectral lines to wobble back and forth—around a much smaller orbit about a hundred times more slowly—only about as fast as the running speed of a desert tortoise. This is because the Earth is 300 times less massive than Jupiter and exerts a much weaker gravitational pull on the Sun. We have the techniques today to detect Jupiter-mass exoplanets. The detection of the much smaller velocity changes produced by exoplanets with masses similar to that of Earth will be extremely challenging. NOAO/AURA

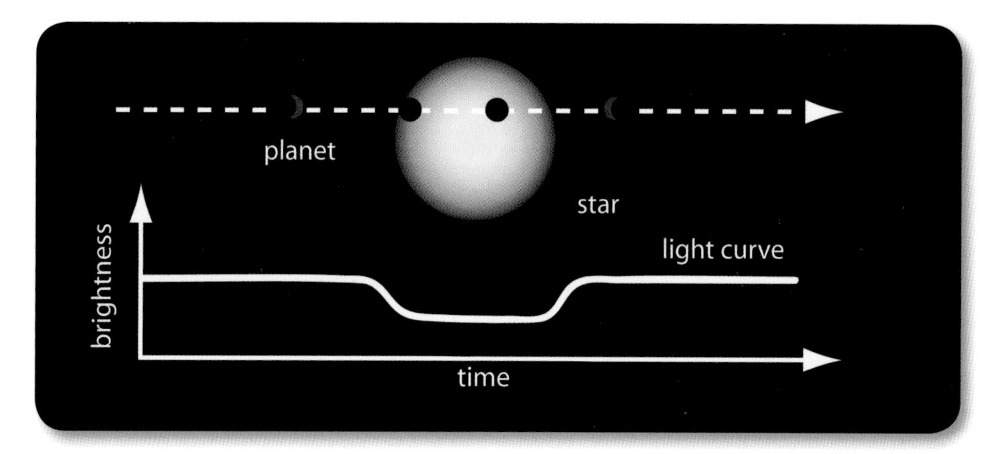

When a planet crosses in front of a star (transits), the star appears to dim slightly. By measuring the depth of the dip in brightness and knowing the size of the star, scientists can determine the size (radius) of the planet. The orbital period of the planet can be determined by measuring the elapsed time between transits. Once the orbital period is known, the average distance of the planet from its sun can also be calculated. NASA Ames

smaller planets at larger distances from their parent stars. For example, NASA plans to fund a new spectrograph for the WIYN Telescope on Kitt Peak that would be able to measure stellar velocities with enough accuracy to detect planets with the mass of Neptune and possibly even planets close to the Earth's mass. The goal is to detect planets that are far enough away from their parent star to be cool enough to sustain life.

Discovery of Exoplanets: Transits

The second technique for searching for exoplanets relies on very precise measurements of the brightness of stars. If a planet transits, that is, passes directly in front of, its parent star, it will block some of the stellar light. The dimming is typically only a few parts in a million. In order to make brightness measurements with this accuracy, it is necessary to get above the Earth's atmosphere. The Kepler satellite, launched in 2009, has discovered more than 1,500 exoplanets, with about 4,000 more candidate planets that need to be confirmed by additional observations.

To the surprise of most astronomers, the first planets to be discovered had masses greater than Jupiter but orbital periods of only a few days—they were even closer to their suns than Mercury is to our Sun. These "hot Jupiters" have very low densities and must, like Jupiter and Saturn, be made mostly of hydrogen and helium. Planets made of these gases could not have formed where they are now. Temperatures so close to a star would be

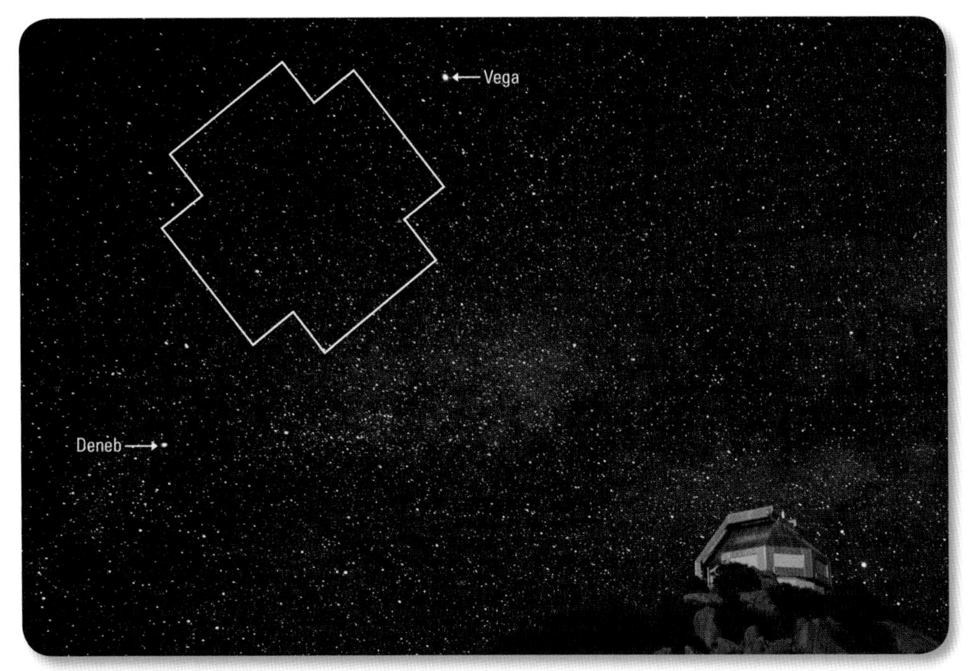

The region in the constellation Cygnus that was observed by the Kepler spacecraft. The squares show the location of the sensors in the Kepler camera. Altogether, the camera had a total of 95 megapixels. Kepler stared at this region of the sky continuously for more than 3 years to look for changes in brightness of more than 150,000 stars in order to discover transiting planets. NOAO/AURA/NSF

very high, and the hydrogen and helium atoms would be flying around so fast that gravity could not trap them and force them to come together to form a planet. Instead, it seems likely that these hot Jupiters formed initially farther away from their suns, just as Jupiter formed in a cold region of our own solar system. As a result of gravitational interactions with other material surrounding their parent stars, these initially cold planets then spiraled into the inner parts of their solar systems and heated up.

The Search for Life

Kepler can also detect planets with masses much lower than Jupiter. Based on the Kepler data, scientists estimate that about 20 percent of Sun-like stars have planets with diameters similar to Earth and that are in the habitable zone—that is, they are at a distance from their parent star where they might have reasonable temperatures. A reasonable temperature is one where liquid water could exist because liquid water is essential for life as we know it. Note the words *might* and *could* in the previous sentences. Being at the right distance from a star is no guarantee of a reasonable planetary temperature. Both Venus and Mars are in the habitable zone as far as their distances from the Sun are concerned. But greenhouse gases make Venus too hot to have retained any water, and all of the liquid water originally on the surface of Mars has evaporated.

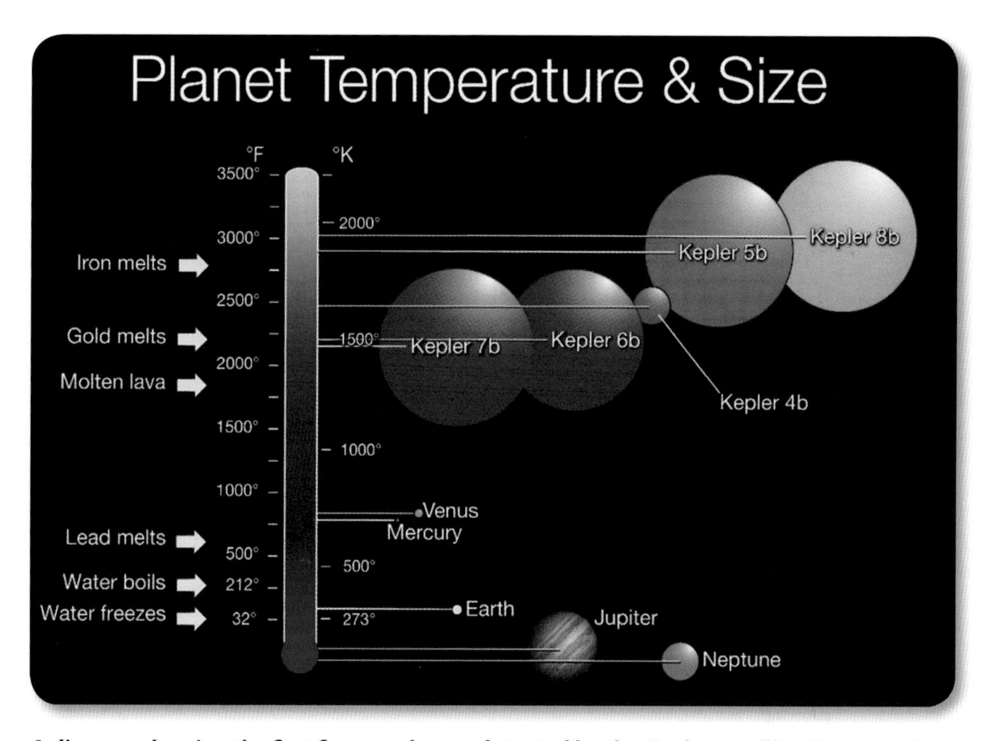

A diagram showing the first five exoplanets detected by the Kepler satellite. These are all examples of "hot Jupiters." The smallest is similar to Neptune in size; the others are even larger than Jupiter. Their surface temperatures are all hotter than molten lava, and some have temperatures high enough to melt gold and iron. Kepler/NASA

Several planets have now been found in the habitable zone, and in July 2015, NASA announced the discovery of Kepler-452b, a planet close to the size of the Earth in the habitable zone around a star that is similar to the Sun. This planet is about 1,400 light-years away from Earth and is too faint for detailed study with current techniques. Nevertheless, an optimistic interpretation of the available statistics about exoplanets suggests that there is likely to be one suitable planet in our own neighborhood—if you consider a distance of 12 light-years from the Sun as nearby, astronomically speaking. In order to search for nearby habitable planets, NASA plans to launch the Transiting Exoplanet Survey Satellite (TESS) in 2017 to search for transiting exoplanets around 500,000 bright stars.

Supposing we do find promising planets in the habitable zone, what can we learn about them? With greatly improved instruments in space, we might be able to image such a planet. The image will cover only a single pixel, but we can learn a lot from a simple image. Such images have been taken of Earth from the distance of Saturn and beyond by spacecraft. The

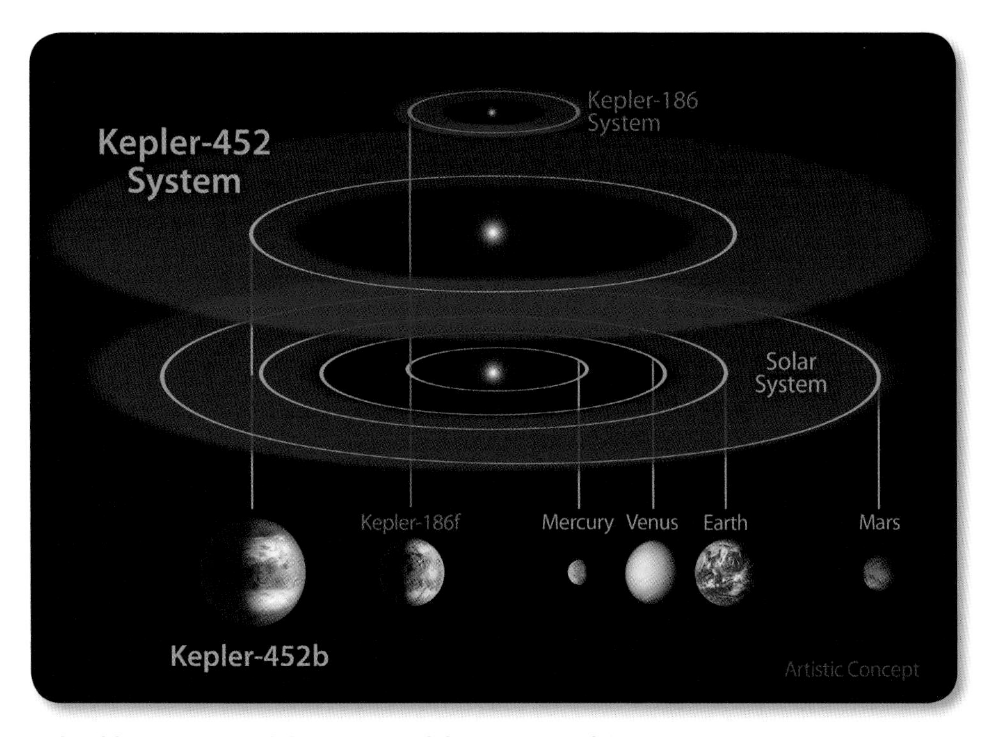

Habitable zones around three stars and the locations of their planets. Hotter stars emit more energy and have larger habitable zones. The star Kepler-186 has a temperature of about 3,800 degrees Kelvin (6,400 degrees Fahrenheit), and its habitable zone is small and very close to the star. Kepler-452 and the Sun both have temperatures of about 5,800 degrees Kelvin (10,000 degrees Fahrenheit). Because Kepler-452 is slightly larger than the Sun and emits more total energy, its habitable zone is larger. The planet Kepler-452b is the first one to be discovered orbiting a star like the Sun that also has a mass similar to that of Earth. Kepler-452b orbits its star every 385 days. It receives about 10 percent more energy from its parent star than the Earth receives from the Sun. Even if additional observations show that Kepler-452b is a rocky planet like the Earth, it may be hot enough to have experienced a runaway greenhouse effect similar to the one that occurred on Venus. NASA/JPL-Caltech/R. Hurt

Earth appears as just a *pale blue dot*, but if we made repeated observations of the color of this dot, we could learn that the Earth has both land and water, because land and water have different colors. We might be able to learn that there are different types of land because the Sahara desert and the rain forest in the Amazon have very different colors, and each covers a large area. By measuring changes in brightness, we could learn the length of the day. As clouds change, they affect the brightness, and so we could tell that the Earth has weather.

EXPLORING ON YOUR OWN

Citizen Science

Citizen science is the term used to describe scientific contributions made by people who are not trained as scientists. Sky surveys and satellites like Kepler produce floods of data, and citizen scientists are playing an important role in making discoveries in these data sets. While much of the analysis is, of course, carried out by computers, human beings have superior pattern recognition capabilities. Computers find what they are told to look for. Humans can find the strange and unexpected.

One of the most successful citizen science projects is Galaxy Zoo, which was initiated in 2007. This project involved classification of 800,000 galaxies, and participants were given six choices to describe the galaxies: Clockwise, anti-clockwise, or edge-on spirals; ellipticals; mergers; and don't know. Tests for a small subset of the galaxies showed that the average classification derived from reports by as many as 80 citizen scientists for each galaxy proved to be as accurate as classifications by professional astronomers. In effect, this is science by crowdsourcing. Several hundred thousand people have contributed to Galaxy Zoo.

Sky surveys also discover *transients*—objects that appear suddenly but were not visible in earlier observations. Professional astronomers work with citizen scientists to determine which ones are good candidates for follow-up. Consensus on classification of a transient as a probable supernova, likely astronomical transient but not a supernova, or artifact can usually be reached within 15 minutes, which makes rapid follow-up possible. The results of human classifications are also used to improve computer programs for classifying transients automatically.

There is another citizen science project called Planet Hunters currently searching for exoplanets. Volunteers are asked to examine the light curves of stars measured by Kepler to look for the dimming of light caused when a planet crosses in front of a star. As of this writing, citizen scientists have discovered fifty planets missed by computer analysis of the same light curves, including the first planet in a system with four suns. This project has been used to evaluate the software used to detect transits automatically and make improvements to it. NASA's TESS satellite to detect exoplanets around nearby stars may have a similar program. The LSST project is already planning ways to engage citizen scientists when its survey begins in around 2022.

Do you want to try a citizen science project? Visit www.zooniverse.org or just type "zooniverse" into a search engine and see what projects are currently available. In addition to astronomy, there are projects in the humanities, climate, and biology. Count penguins, transcribe ship logs, monitor California condors—there are many choices, and each comes with an easy tutorial so that you can see whether that project interests you. Check it out! ●

We know, of course, what the Earth is actually like, so it is easier to interpret observations of color and brightness. An exoplanet may bear no resemblance to the Earth, and so interpreting color and brightness changes may be more challenging. Scientists, however, are already preparing for the day, probably a few decades in the future, that we may be able to make such observations. For example, scientists are cataloging the colors of microorganisms, including some found in the most extreme environments on Earth. If one of these microorganisms is a dominant life form on an exoplanet, then it might be detected by measurements of the exoplanet's color.

Earth as seen by the Cassini spacecraft orbiting Saturn. The Earth, which is 898 million miles away in this image, appears as a blue dot below the rings; in the original image, the Moon can be seen as a fainter protrusion off its right side. In this image, Saturn appears dark because we are looking at its night side, which is not illuminated by the Sun. The edge of Saturn looks bright except where the shadows of the rings block the sunlight. NASA/JPL-Caltech/Space Science Institute

To move beyond potential habitability to evidence of life itself, it will be necessary to search for biosignatures. Let's assume that any form of life would use chemical reactions to extract energy, store it, and release gases as a by-product of metabolism. Proof of life would be the discovery of a combination of gases in the atmosphere that require ongoing production for their existence. An example on Earth is that methane and oxygen are both present in the atmosphere. The two combine to produce water and carbon dioxide. Oxygen and methane coexist only because they are continually replenished by biological processes. On Earth, we know that methane is produced as a by-product of metabolism in animals and oxygen by photosynthesis.

While detection of oxygen and methane in the atmosphere of another planet would be convincing evidence of life, we have to entertain the possibility that life may not always resemble life on Earth. Cells are the basic component of life on Earth and are probably the simplest, essential unit of all forms of life. Inside a cell, chemical reactions take place that use nutrients and energy to enable the cell to grow and reproduce. A membrane surrounds the cell and protects it from the external environment, which might disrupt the reactions needed to sustain life. On Earth, cell membranes can provide this protective function only in the presence of water, and that is why the so-called habitable zone is one with temperatures suitable for the existence of liquid water. Researchers are now beginning to explore the possibility that a different type of cell membrane could function in an environment where, for example, methane is a liquid. Could life have formed in the methane lakes on Titan?

Is there life, even primitive life, elsewhere in the Universe? Or are we alone? Looking for answers to these questions is a main focus of the NASA science program. The search is on!

Formation of Stars and Planets

For most of the twentieth century, the formation of stars and planets was the domain of theorists because these steps in the process were hidden in dark clouds filled with dust. Also, the formation of individual stars happened on such a small scale that we could not observe the details. Now with sensitive infrared detectors that can see through the dust and the high spatial resolution offered by the Hubble Space Telescope and the Atacama Large Millimeter Array in Chile, we can actually observe the process of star formation and validate our theories. We no longer need to always rely on an artist's conception! ●

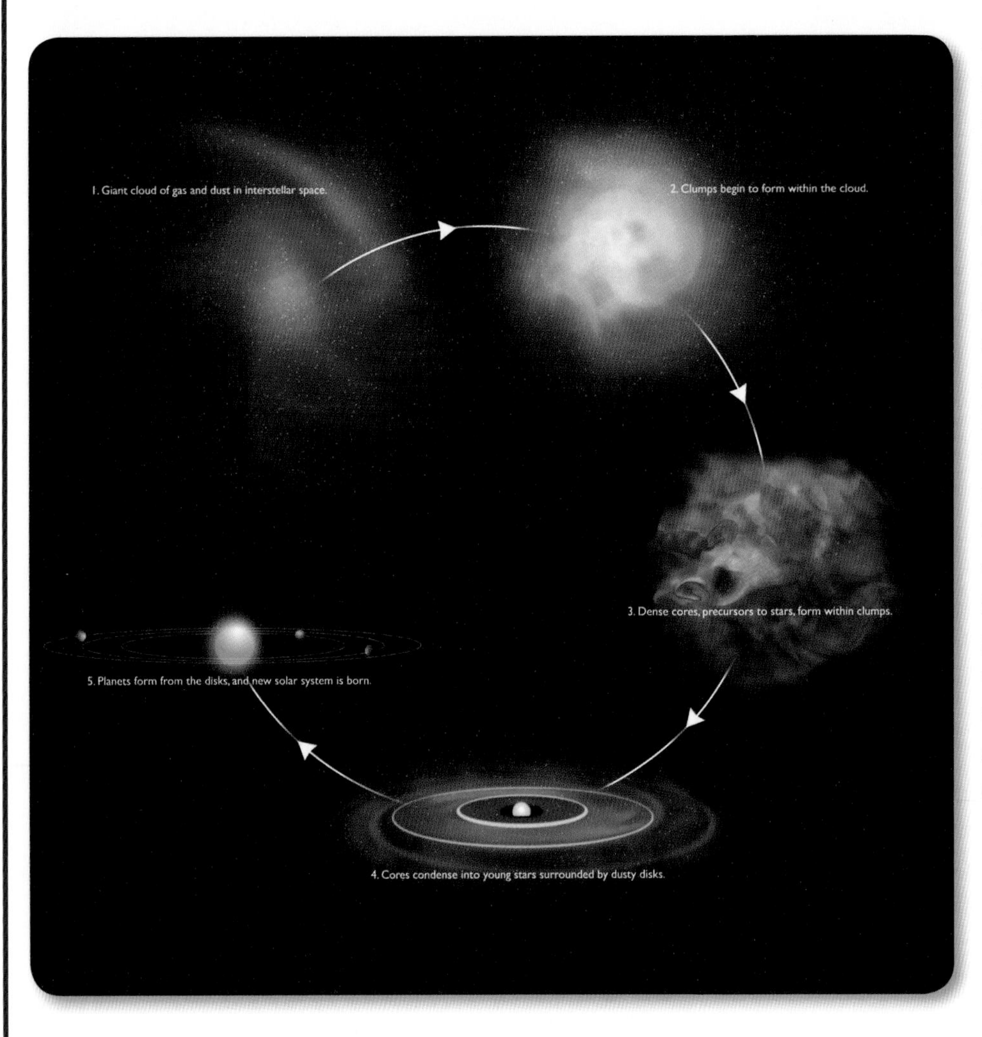

◀ **An artist's conception of the stages of star formation.** The process begins in clouds of dust and gas. Denser regions within the clouds begin to collapse due to gravity and form large clumps of gas and dust. These clumps then fragment into still denser cores, which collapse to form stars. Newly formed stars are surrounded by a disk of dust and gas. Over a few million years, the dust particles collide and stick together to form sand, then pebbles, larger rocks, and ultimately planets.

NRAO/AUI/NSF

◀ **Visible and infrared views of star formation in the Trifid Nebula.** A composite of two visible-light images of the Trifid taken with NASA's Hubble Space Telescope and at Kitt Peak (left) shows a murky cloud lined with dark trails of dust. Newly forming protostars are indicated with arrows in the infrared Spitzer Space Telescope picture (right). These very young stars cannot be seen in the visible-light picture (left). NASA/JPL-Caltech/J. Rho (SSC/Caltech)

▼ **A star-forming cloud in the constellation of Orion.** This composite image combines Hubble Space Telescope observations in visible light (color-coded in blue and green) with Spitzer Space Telescope images in the infrared (color-coded in orange and red). Stars appear as dots sprinkled through the image. Orange-yellow dots are infant stars shrouded in dust and gas and can be observed only in the infrared. This nebula is about 1,500 light-years away from Earth and can be seen in Orion's sword. NASA/JPL-Caltech/T. Megeath (University of Toledo) and M. Robberto (STScI)

▲ **Image of the young star HL Tau and its protoplanetary disk taken with the ALMA radio telescopes located in Chile.** Note the concentric rings and gaps. As newly forming planets circle the central star, they accumulate the dust particles along their paths and sweep their orbits clean. HL Tau is only about a million years old, and this observation shows that planets form fairly quickly, astronomically speaking. HL Tau is a Sun-like star and will live for billions of years. It is located approximately 450 light-years from Earth in the constellation Taurus. ALMA (NRAO/ESO/NAOJ); C. Brogan, B. Saxton (NRAO/AUI/NSF)

A CLOSER LOOK

Colors of Astronomical Images

The illustrations in this book show a variety of ways that color is used to convey scientific information. Infrared images are often color-coded to show temperature (see Chapter 2). Images of planetary surfaces are sometimes shown with exaggerated colors to emphasize the different mineral content of various geological surfaces (see Chapter 4). Other images are taken through filters that transmit the emission only of specific elements, such as the red emission line of hydrogen (see Chapter 2). Maps produced by radio interferometers at wavelengths that the human eye cannot see at all are converted into images that we can see (e.g., the ALMA image of a protoplanetary disk on the preceding page).

The most stunning images of all have been produced from images taken with the Hubble Space Telescope. Here the colors often convey scientific information, but artistic judgment is involved as well. Let's see how the Hubble images are produced.

Think first about how a digital camera or the camera in your smartphone works. There is an electronic detector that senses incoming light and records its intensity. Separate pixels record red, blue, or green light. Software in the camera then combines these separate measurements to produce a color image. Different intensities of R, G, and B can combine to produce any arbitrary color. Your eye works in a similar way. Cones in the eye are sensitive to long, medium, or short wavelength light (roughly but not exactly corresponding to RGB), and the brain combines them so that we can see color.

In the case of the Hubble cameras, each pixel in the sensors can detect the same broad range of wavelengths. The cameras are equipped with filters that transmit specific colors or wavelengths of light. Thus, in order to make an RGB image, it is necessary to take three different pictures—one through each of the red, green, and blue filters. All of the images are initially grayscale. Color is added to each image and computer processing then combines the separate images to produce the final version for publication. In practice, the Hubble Space Telescope is equipped with many different filters. Some transmit only a narrow portion of the electromagnetic spectrum, such as the red line of hydrogen. ●

▲ This image shows emission from sulfur. It was colored red for the final image.

▲ This image shows emission from hydrogen. It was colored green for the final image.

▲ This image shows emission from oxygen. It was colored blue for the final image.

▲ The final image of the Pillars of Creation.

Individual grayscale Hubble images and final color image of famous Pillars of Creation. Stars are being born deep inside the pillars, which are made of cold hydrogen gas laced with dust. The pillars are part of a small region of the Eagle Nebula, which was shown in the first image in Chapter 1. This vast star-forming region is 6,500 light-years from Earth. Z. Levay and NASA, ESA, and the Hubble Heritage Team (STScI/AURA)

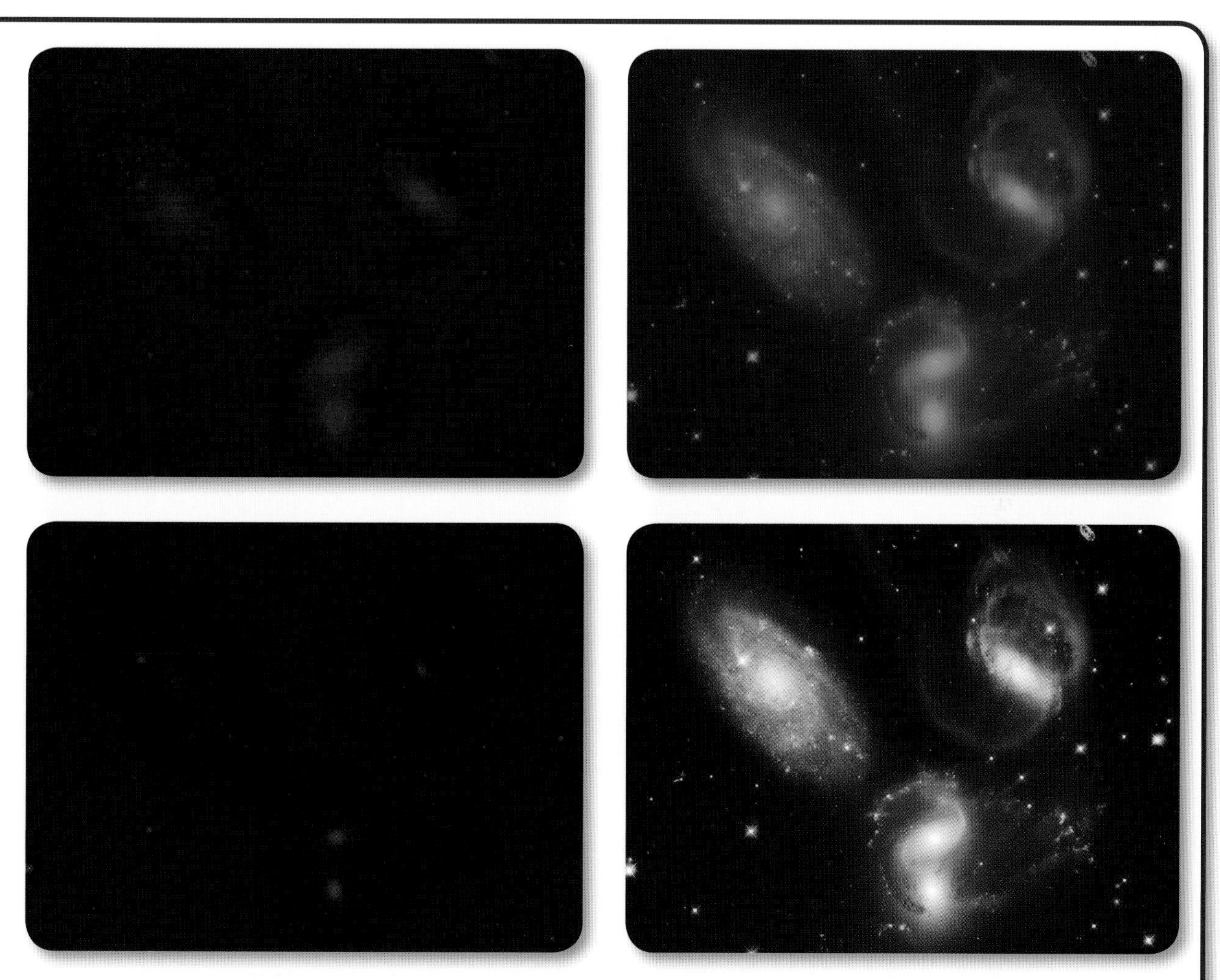

▲ **Four of the members of Stephan's Quintet, a group of galaxies that appear close together in the sky.** Shown are separate RGB images, where color has been added to the original grayscale images. The color images tell astronomers a lot about these galaxies that is not easily seen in grayscale images. NGC 7319, at top right, has distinct spiral arms that curl nearly 180 degrees back to the central part of the galaxy. Continuing clockwise, the next galaxy appears to have two cores, but it is actually two galaxies, NGC 7318A and NGC 7318B, that are colliding with each other. Around these galaxies are young, bright blue star clusters whose member stars are less than 10 million years old. Hot, luminous blue stars have lifetimes measured in millions of years, and so blue color is a sign of recent star formation. (The Sun, in contrast, is much cooler, appears yellow, and has a lifetime measured in billions of years.) The pink color is produced by hydrogen gas. It takes hot blue stars to make hydrogen glow, so we know that stars are also still forming in the pinkish clouds. The galaxy at the upper left, NGC 7320, is also forming stars as shown by the blue and pink dots. This galaxy is much closer than the others. NGC 7320 is 40 million light-years from Earth. The three other galaxies shown here are at a distance of 290 million light-years. Z. Levay and NASA, ESA, and the Hubble SM4 ERO Team

Two colliding galaxies. We now know that over the lifetime of the Universe, galaxies grow by colliding and merging with other galaxies. This pair of galaxies is named the Antennae, and they began to interact a few hundred million years ago. The orange blobs to the left and right of the center of the image are the bright cores of the original galaxies and consist mainly of old stars. The collision compresses the gas and dust in the galaxies and causes new stars to form. The regions of recent star formation contain hot blue stars surrounded by glowing hydrogen gas, which is shown in pink. This image may give us a preview of what will happen when our own Milky Way collides with the nearby Andromeda Galaxy in a few billion years.

NASA, ESA, and the Hubble Heritage Team (STScI, AURA, ESA /Hub)

6 Evolution of Stars and Galaxies

Astronomers working in the last part of the twentieth century and the early part of the twenty-first often refer to this era as the Golden Age of Astronomy. New technologies have enabled discoveries in our own lifetimes at a pace that continues to amaze even professional astronomers. In 1610, Galileo's observations with a new technology—a small telescope—completely changed our view of the solar system by showing that the Earth could not be at its center. In just the same way, the dramatic discoveries of the twentieth century have enabled researchers to describe the entire history of the Universe since it began to expand. (The theory that the Universe initially had a very high temperature and was at a very high density and began to expand rapidly is called the big bang theory.)

At the beginning of the twentieth century, we did not even know what stars are made of. We now know that stars are made mostly of hydrogen and helium and that the Sun and most stars produce energy through fusion reactions that convert hydrogen to helium (similar to what happens in hydrogen bombs). The helium atom is slightly less massive than the four hydrogen atoms that fused together to form it, and that lost mass is converted to energy. Recall Einstein's equation $E = mc^2$, which says that energy (E) and mass (m) are equivalent. The constant that defines the relationship between energy and mass is c, the speed of light. Because the speed of light is a very large number (186,000 miles per second), small amounts of mass can be converted to very large amounts of energy.

With this and other information, we can trace the entire life story of a star from birth to death. For example, we can calculate how long the Sun will continue to shine at its current rate. We know how much hydrogen it has, and we know that in order to produce the amount of energy we see from the Sun, four hydrogen atoms must be converted to helium 10^{37} times every second (that is 10 multiplied by itself 37 times—an unimaginably large number). This sounds like a lot of hydrogen, but the Sun is very massive. It has already been shining for about 4.8 billion years and has enough hydrogen to continue to shine for another 5 billion years or so. Some stars produce energy at such a prodigious rate that they exhaust their nuclear fuel in only a few million years.

Other breakthroughs relate to the study of galaxies. At the beginning of the twentieth century, we knew only that there are many stars near the Sun, but we didn't know those stars were organized into a galaxy or that other galaxies exist. Now we know that the Milky Way, the Sun's home galaxy, contains about 300 billion stars and is a little over 100,000 light-years across. (Astronomers deal with such large distances that they use light-years rather than miles as their distance measurement. One light-year is the distance that light, which has a speed of 186,000 miles per second, travels in one year; one light-year is equal to about 6 trillion miles.)

We also now know that the Milky Way Galaxy is not the only galaxy. Astronomers estimate that there are 125 billion galaxies in the Universe. Measurements show that all of these galaxies are moving away from each other. In other words, the space between galaxies is getting larger. If we think of the Universe as a movie, we can run it backward and imagine a time when all of the galaxies were infinitely close together. This time occurred, according to the best modern estimates, 13.8 billion years ago, which we say is the age of the Universe. The expansion of the Universe thus began 13.8 billion years ago. Because the initial expansion was very rapid, scientists refer to the beginning of the expansion as the big bang.

Combining information about stars and the age of the Universe, we learn another important fact. Most stars that we observe are younger than the Universe—that is, they formed long after the expansion began. For example, the Universe was already 9 billion years old before our Sun was formed, and we have found stars that are only a few hundred thousand years old. This observation means that the contents of the Universe and the stars within it change with time. Some changes, such as the slow evolution of the Sun, take billions of years. Others, such as the formation of planets, take no more than a few million. And some rare events, including the explosive deaths of stars, occur in days.

Evolution of Stars

Let's try an analogy to understand the challenge astronomers faced when trying to determine how stars evolve. Imagine you are an alien visiting Earth for a single day, and your scientific assignment is to determine how human beings change with time. You could try watching one individual human for 24 hours, but you likely wouldn't see any change at all. Alternatively, you could survey a lot of humans and record their characteristics. Some are large, some small; some have hair and some don't. Humans come in different colors. Which characteristics are clues to the life cycle of a human? Unless the alien has a good understanding of biological processes, it would have a hard time sorting out that height and weight are good clues to age, for example, at least during the first 15 years of human life.

NGC 1073, a spiral galaxy that resembles the Milky Way. Obviously, we cannot fly outside of the Milky Way to view our Galaxy at a distance. Instead, astronomers look for other galaxies that resemble our own. NGC 1073 has a bright nucleus with a bright bar of stars extending from it. Our own Galaxy also has a bar, but it is less prominent than this one. Both galaxies have spiral arms that contain bright blue stars, clouds of glowing hydrogen gas (shown in red), and dark dust lanes. Light takes about 55 million years to reach us from NGC 1073, which has a diameter of about 80,000 light-years. NASA & ESA

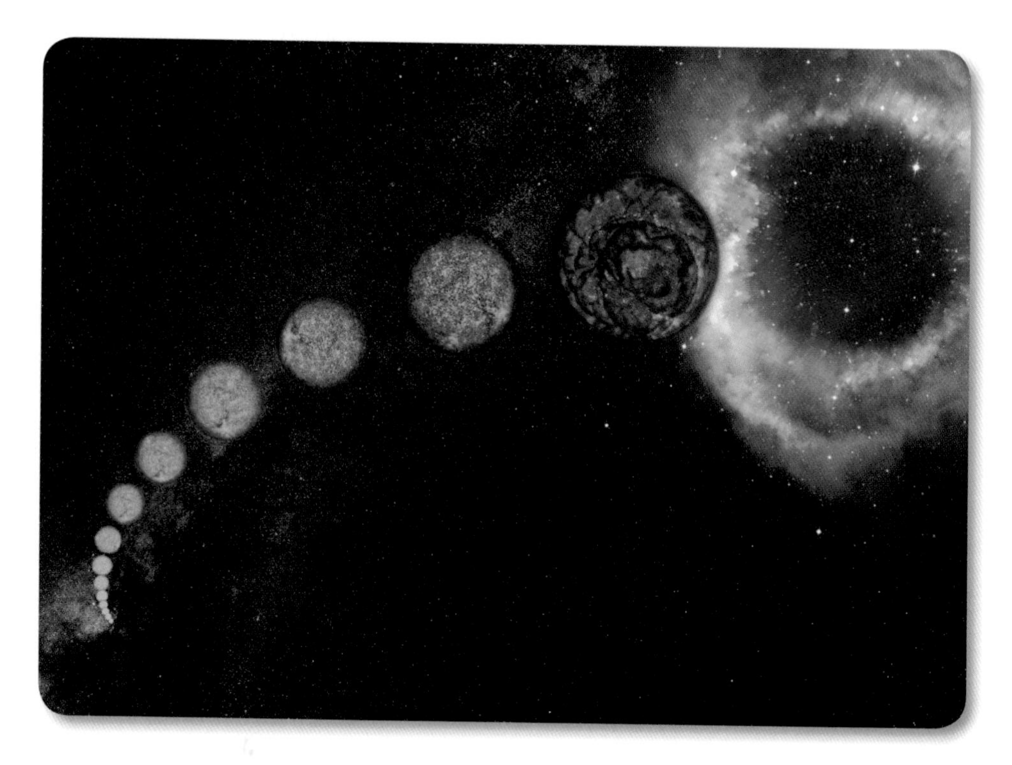

Evolution of the Sun from birth to death. About 10 billion years after it formed and about 5 billion years from now, the Sun will expand to become a red giant. It will shed some of its outer layers to form what astronomers call a planetary nebula before it completely exhausts its fuel and collapses to become a tiny white dwarf. ESO/S. Steinhöfel

Astronomers had to approach stellar evolution in much the same way. First they surveyed stars and determined that stars come in a range of sizes: from about a tenth the mass of the Sun up to 150 times the mass of the Sun. The most massive stars emit the most energy. Stars also have different temperatures.

Astronomers have a big advantage, however, because they know the laws of physics. Observations show that gravity behaves in the same way everywhere in the Universe. We know how much pressure the hot gases inside a star must exert to resist the pull of gravity, which would otherwise cause the star to collapse. We know how much energy is produced by fusion reactions and how that energy is transported from the center of the star to the surface, where we can measure it. Astronomers use this information to build model stars that match the real stars that we observe. And then we go a step further, and calculate how that model changes if we let the nuclear reactions continue for a period of time. In this way, we can calculate how a star like the Sun evolves from the time it is formed until the time it runs out of nuclear fuel.

Let's take a detailed look at the evolution of the Sun. Like all stars, the Sun formed initially in a cloud of dust and gas. The Sun will spend most of its life—about 10 billion years—generating energy by fusing hydrogen to form helium. During this time, its energy

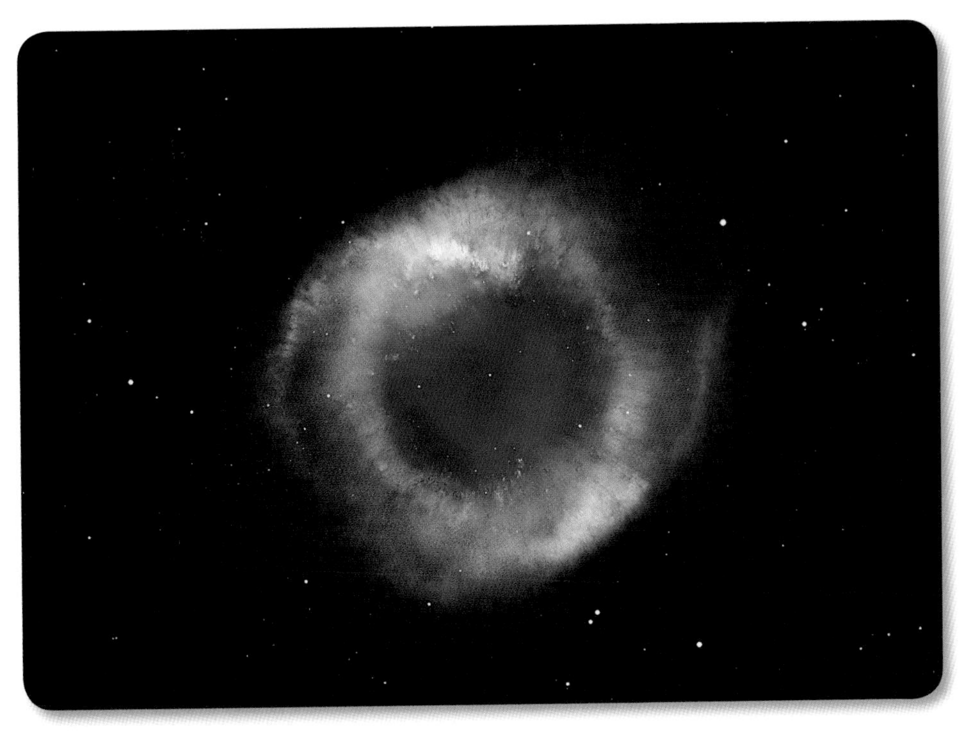

A planetary nebula. This image of the Helix Nebula is a blend of nine images from the Hubble Space Telescope with a wide-field-of-view image taken at Kitt Peak National Observatory. Astronomers calculate that several billion years from now, the Sun will eject gas from its outer layers and this gas will form a planetary nebula similar to the Helix Nebula. The Sun will then become a faint white dwarf. In this image the white dot at the center of the ring is a white dwarf. This nebula is about 700 light-years from Earth. The colors in this image correspond to glowing oxygen (blue), and hydrogen and nitrogen (red). NASA, NOAO, ESA, the Hubble Helix Nebula Team, M. Meixner (STScI), and T.A. Rector (NRAO)

output increases slowly. The Sun emitted about 30 percent less energy when it first formed 4.6 billion years ago than it does now. The Sun's hydrogen fuel at the core will finally be exhausted in about 4.8 billion years, when the Sun will be 67 percent more luminous than at present. Thereafter, the Sun will expand, its surface temperature will decrease, and it will become a cool, red giant star with a diameter that will extend beyond the Earth's current orbit. Long before then, however, increased heating by the Sun will evaporate enough of the water in the oceans to cause a runaway greenhouse effect—just as we think happened already on Venus, making that planet uninhabitable. Fortunately, calculations show that this won't happen for another billion years or so.

After it becomes a red giant, the Sun will shed some of its outer layers and form a *planetary nebula*. A planetary nebula has nothing to do with planets but was given this name because when seen through a telescope these nebulae resemble disks, just as the planets do. Eventually, the Sun will completely run out of nuclear fuel, and it will shrink to become a white dwarf. It will then be about the diameter of the Earth but it will be very dense because it will still have most of its mass. A matchbox filled with a chunk of white dwarf material would weigh 275 tons. Since the white dwarf has no source of energy, it will gradually cool off, cease to emit any energy, and become an invisible black dwarf.

How to Find a Black Hole

The idea that black holes might exist goes back more than 200 years, but proof that they actually do exist came only in the last half of the twentieth century.

By definition, a black hole is a region of space where the force of gravity is so strong that nothing can escape from it—not particles of matter and not even light.

In the early eighteenth century, scientists knew that light didn't travel at an infinite speed, and their estimates of light's speed were close to the modern value of 186,000 miles per second. Scientists also knew that the velocity required to escape the gravitational pull of, for example, a planet was higher for planets with higher masses. As a specific example, if a rocket is to escape the Earth completely, it must be launched at a speed of 11 kilometers per second (25,000 miles per hour). The escape velocity from Jupiter, which is 300 times more massive than the Earth, is 60 kilometers per second (134,000 miles per hour).

Putting these two ideas together, in 1783 John Michell wrote a paper speculating that there might be an object with such a strong gravitational pull that even light could not escape. The term *black hole* was first applied to such an object in 1969. Einstein's general relativity theory is required to explain what happens if gravity is actually this strong.

The boundary that marks the point of no escape is called the *event horizon*. Anything within the event horizon is forever trapped and cannot communicate in any way with the world outside. How big is the event horizon? For the Earth to become a black hole—which can't happen—it would have to be compressed to fit within a sphere with a radius of 1 centimeter—about the size of a grape.

There is a myth that black holes are monsters that go about gobbling up everything with their gravity. Actually, very strange things happen only very close to a black hole. If the Sun were to become a black hole, with all of its matter crammed inside a region with a radius of about 3 kilometers (less than 2 miles), we here at the Earth would notice the loss of heat, but the Earth itself would continue to sail serenely in the same orbit.

▲ **An artist's conception of the binary system Cygnus X-1, which provided the first evidence that black holes really do exist.** The black hole in this binary contains about five times the mass of the Sun, squeezed into a tiny sphere only a few miles in diameter. Because of its density, it possesses an enormous gravitational field, which is pulling matter away from its companion star, HDE 226868. The companion is a massive star, known as a blue supergiant, with a surface temperature of 31,000 K. As the gas spirals toward the black hole, it is heated even further and gives off X-rays and gamma rays. NASA, ESA, Martin Kornmesser; ESA/Hubble

What kinds of objects might turn into black holes? To astronomers, the most likely possibility seemed to be massive stars that have run out of energy. Calculations had shown that once their energy is exhausted, stars significantly more massive than the Sun cannot resist the force of gravity and will collapse to an infinitely small volume. Think of a balloon that suddenly loses all its air.

How would we find such an object? We can't see it, so all we can detect is its gravitational force on a nearby star. So we are looking for a binary star where one star is visible and has an invisible companion with a mass greater than about three times the mass of the Sun (a star with a mass lower than this does not collapse all the way to a black hole). Invisibility is not enough. We must make sure that the unseen companion also has a very small radius. Let's suppose that mass is being shed from the star we can see and is falling toward the black hole. Calculations show that this mass will swirl around the black hole before falling in, much as water swirls around the drain in a bathtub. Unlike water, however, the gas falling toward a black hole reaches a very high speed and becomes so hot that it emits X-rays. The first binary that met all these criteria—massive invisible companion and X-ray emission—was Cygnus X-1. Based on observations in the early 1970s, this system provided the first observational evidence that black holes really do exist. Many similar systems are now known.

In the last decade, observations have shown that most galaxies harbor much more massive black holes in their centers. The masses of these black holes range from a few million to several billion times the mass of the Sun and are correlated with the masses of their host galaxies—more massive galaxies host central black holes with higher masses. This correlation suggests that the formation of the central black hole and the formation of the galaxy itself are closely linked. Trying to understand that linkage is an area of very active research. ●

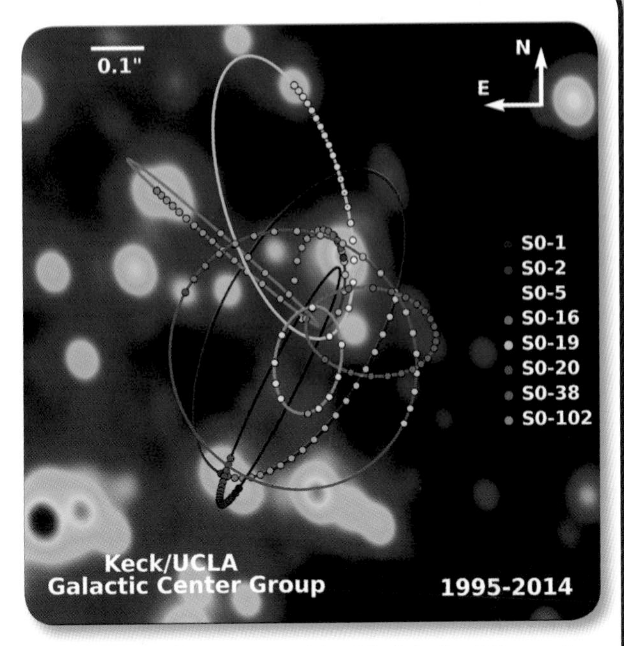

▲ **The orbits of stars around the black hole at the center of our Galaxy.** In the background, the central portion of an image of this region is displayed. The motions of the stars along their orbits are plotted as colored dots. These orbits make it possible to estimate the mass of the object at the center of the Galaxy. The mass controlling these orbits must fit inside the orbits themselves. That means that the mass must fit in a volume with a diameter less than the diameter of the orbit of Uranus. The mass itself must be a little over 4 million times the mass of the Sun. Only a black hole meets these two requirements: huge mass crammed into a very small space.
This image was created by Professor Andrea Ghez and her research team at UCLA from data sets obtained with the W. M. Keck Telescopes.

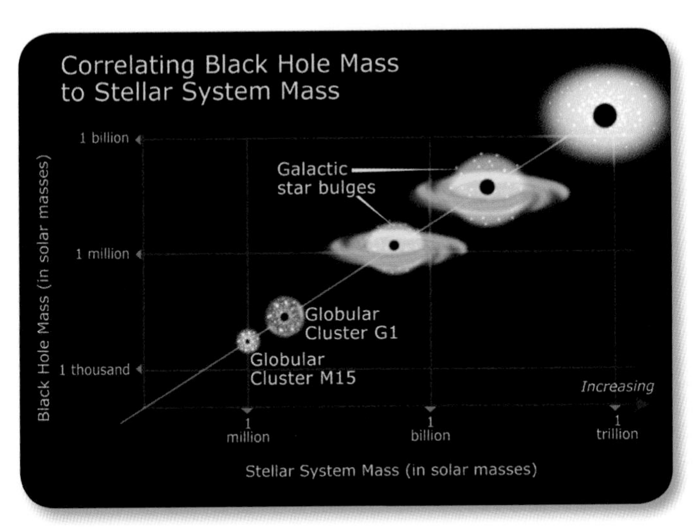

◀ **Most galaxies have black holes lurking in their centers.** The more massive the galaxy, the more massive its central black hole. The horizontal axis indicates the range of masses of galaxies. The vertical axis indicates the range of black hole masses.
K. Cordes, S. Brown (STScI) from data sets obtained with the W. M. Keck Telescopes.

All stars begin their lives in clouds of dust and gas, but they do not all end their lives in the same way. Stars with masses of about 25–150 times the mass of the Sun shed a lot of mass and end their lives in a giant explosion called a supernova. This explosion blows off the outer layers of the star and leaves behind a black hole (see A Closer Look on pages 118-119). The Web has information about the life cycles of stars of different masses. Try using search terms like "stellar evolution" or word combinations like "supernova black hole."

Evolution of the Universe

We cannot observe the evolution of a single star in a human lifetime, and so we must rely on computer models to tell us how stars evolve. Given that, it may seem odd that we can directly *observe* the evolution of the Universe as a whole.

Because light does not travel instantaneously from one place to another, the Universe is, in fact, a kind of time machine. As we know, light travels at a speed of 186,000 miles per second, which is very fast, but not infinitely fast. For example, the Sun is 93,000,000 miles away. At the Earth, we see light that left the Sun 500 seconds or a little over 8 minutes ago. If the Sun were to suddenly cease to shine (which it cannot do!) we wouldn't know it until a little over 8 minutes later. Proxima Centauri, a companion to Alpha Centauri, is the nearest star at a distance of 4.24 light-years, which is equivalent to 25 trillion miles. What this means is that the light we see from this star left it 4.24 years ago, and we are seeing it as it was then.

We know from measurements of the rate of expansion of the Universe that the expansion began 13.8 billion years ago. Therefore, if we could observe galaxies that are at a distance of more than 13 billion light-years, we would be seeing them very close to the beginning of time. The challenge is that the farther away an object is, the fainter it will look to us. So one reason for building ever-larger telescopes both in space and on the ground is to observe extremely faint galaxies to see what they were like when the Universe was young.

Scientists can now describe all of the phases of evolution of the Universe. Very shortly after the beginning, which we call the big bang, the Universe underwent a period of very rapid expansion. We haven't actually observed this initial expansion and we don't know what caused it, but rapid expansion (astronomers call it inflation) is needed to explain later events that we have seen.

After inflation, the Universe continued to expand but much more slowly. At this time, the Universe was extremely hot and filled with energetic photons. In fact, it was so hot that no atoms could survive. When the Universe was about a millionth of a second old, it had cooled enough that protons and neutrons began to appear. These particles were formed by the collision of energetic photons according to Einstein's equation $E = mc^2$. We have already

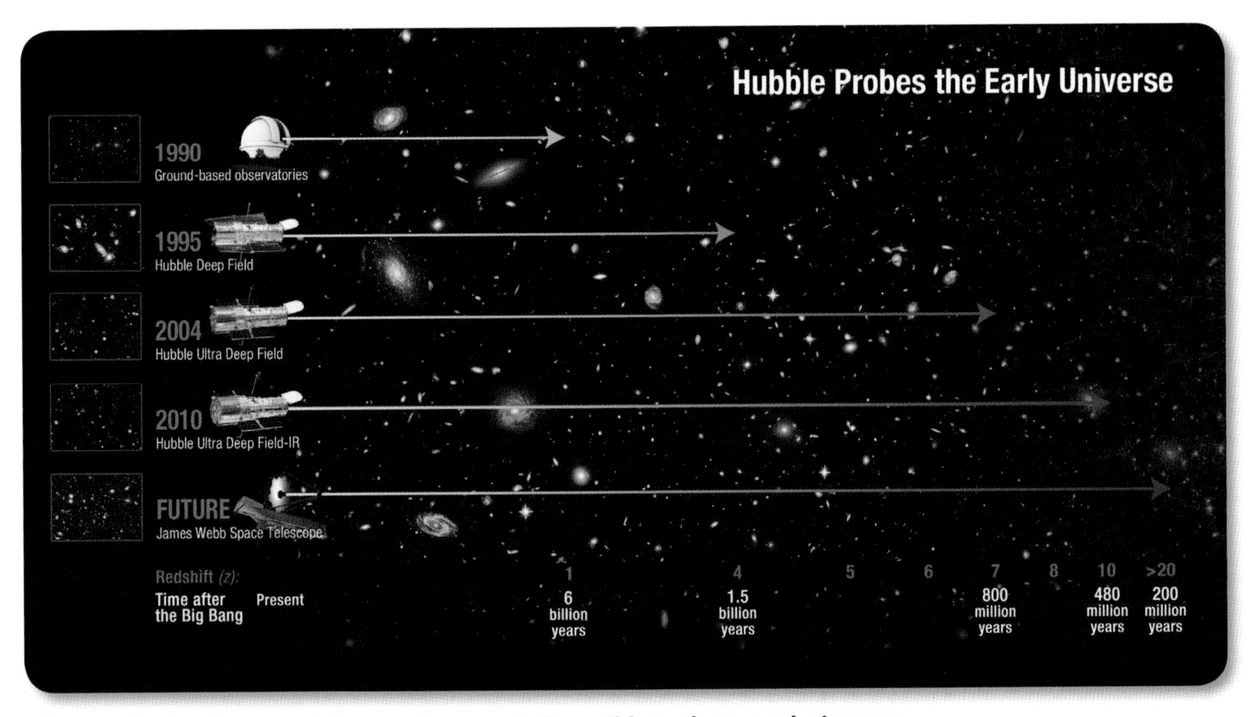

Diagram showing how new telescopes have made it possible to observe galaxies ever closer to the beginning of the Universe. The horizontal axis shows the ages, measured in time since the big bang, of the youngest galaxies that were observed in the years 1990 to 2010. Since the Universe is expanding, the spectra of galaxies are all shifted to the red. The horizontal axis also gives the corresponding value of z, which is a quantitative measure of the redshift used by astronomers. [Mathematically, $1 + z =$ (observed wavelength)/(wavelength emitted by a non-moving source).] The James Webb Space Telescope, which is scheduled for launch in 2018, is larger than Hubble and should allow us to see galaxies even closer to the time that they first started to form. ILLUSTRATION CREDIT: NASA, ESA, and A. Feild (STScI) SCIENCE CREDIT: NASA, ESA, G. Illingworth (University of California, Santa Cruz), R. Bouwens (University of California, Santa Cruz, and Leiden University), and the HUDF09 Team

learned that matter can be converted to energy in the interiors of stars and in hydrogen bombs. But the equation works in the other direction, too. If two photons with sufficient energy collide, they can produce matter. In the extremely hot early Universe, photons had enough energy to make the protons and neutrons we see today. By about a hundredth of a second after the big bang, the temperature had dropped to about 100 billion degrees K, and the colliding photons were no longer energetic enough to produce protons and neutrons.

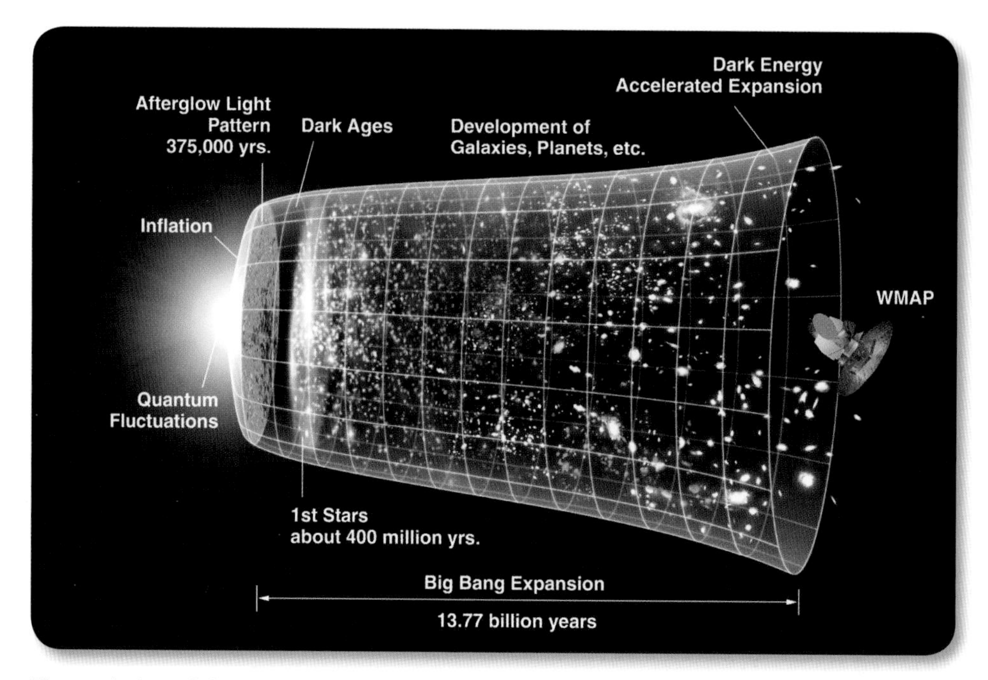

The evolution of the Universe over 13.77 billion years. The size of the Universe is illustrated by the vertical height of the image. The far left depicts a period of exponential growth that scientists call inflation. For the next several billion years, the expansion continued at a more sedate pace. However, during the most recent few billion years, the expansion of the Universe has begun to speed up—evidence for the existence of some new source of energy, which scientists call dark energy. The afterglow light from the hot early Universe that was emitted about 375,000 years after inflation has traversed the Universe largely unimpeded since then. The Wilkinson Microwave Anisotropy Probe (WMAP), shown on the right in the diagram, was one of the space experiments used to observe this afterglow. NASA/WMAP Science Team

By about 3 minutes after the big bang, the Universe had cooled enough that deuterium and helium nuclei could form. Fusion stopped after the first few minutes, but the Universe remained hot and opaque, like the interior of the Sun. Electrons still moved freely, not attached to any atoms, and electrons are particularly effective at scattering photons and changing their direction so that they cannot travel very far. It was only at an age of about 375,000 years, when the Universe had cooled to about 3,000 degrees K, that electrons could combine with the nuclei to form true atoms. With no free electrons to scatter them, photons could travel very great distances—such great distances that scientists on Earth can detect them after a journey of almost 13.8 billion light-years through a now transparent Universe.

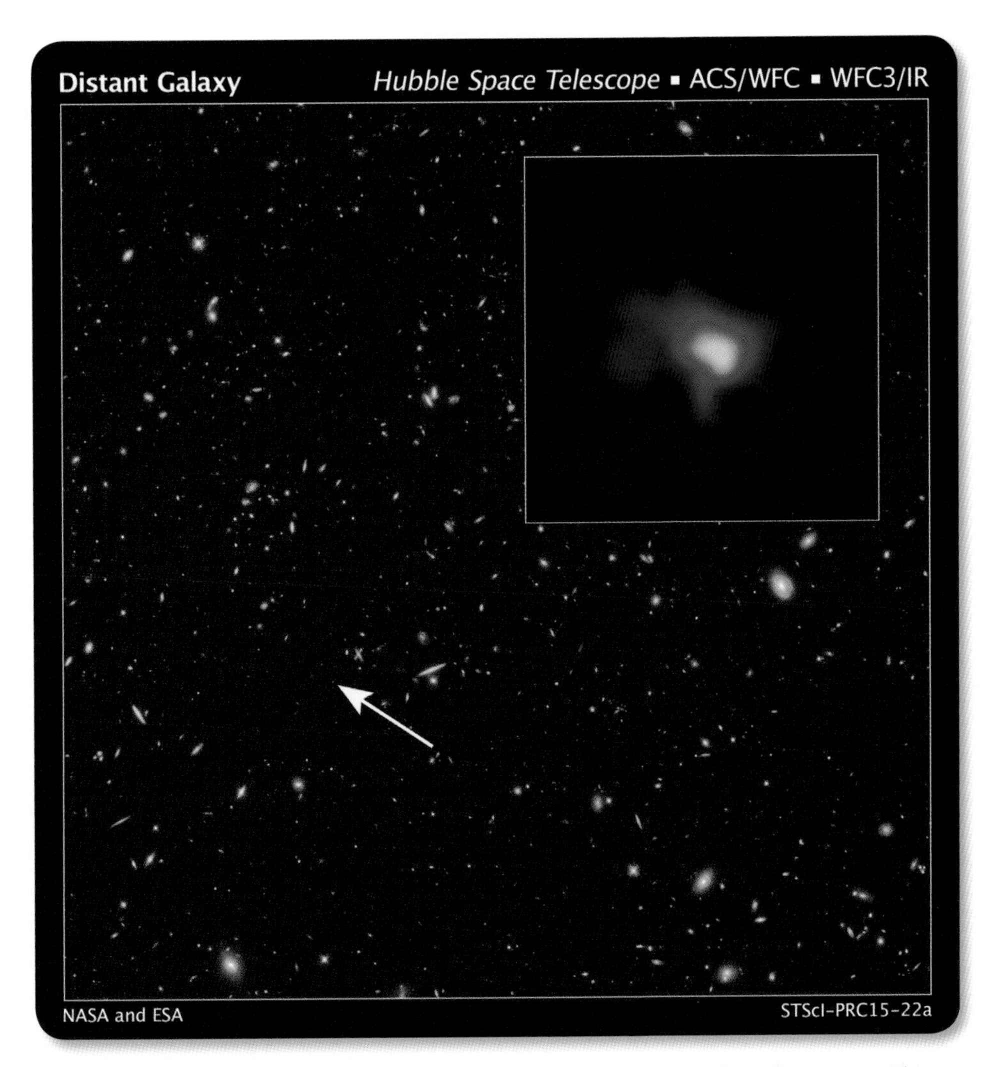

Hubble Space Telescope image of the most distant galaxy observed as of May 2014. This galaxy existed over 13 billion years ago—when the Universe was only 5 percent as old as it is now. This galaxy has only about 15 percent of the mass of our own Milky Way Galaxy but is forming stars at a rate that is 80 times higher than the rate of star formation in our Galaxy. The blue color is an indicator that luminous blue stars are being formed at a very high rate. Star formation could occur at a higher rate then than now, because in today's Universe, much of the gas is already locked up in stars. More gas was available 13 billion years ago. NASA, ESA, P. Oesch and I. Momcheva (Yale University), and the 3D-HST and HUDF09/XDF Teams

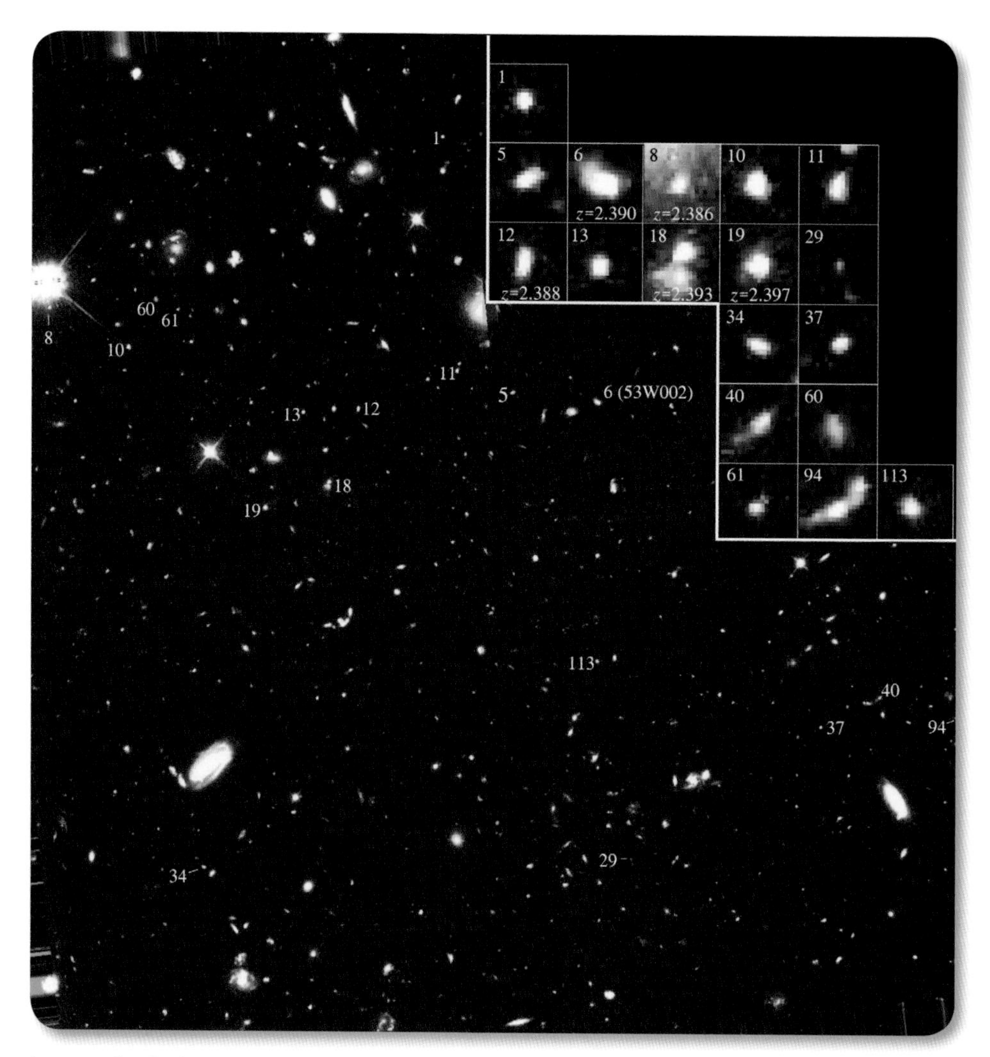

Image of galaxies in the first stage of their evolution. In this Hubble image, astronomers identified 18 bluish objects at a distance of 11 billion light-years. They are scattered across an area of about 2 million light-years. Each object contains only about a billion stars, and is only about 2,000 light-years across—much smaller than typical galaxies in today's Universe. The Milky Way Galaxy is 100,000 light-years across and contains about 300 billion stars. Because these 18 objects are fairly close together, it may be that, over the next several billion years, they will collide and merge to form a true galaxy. Rogier Windhorst (Arizona State University) and NASA

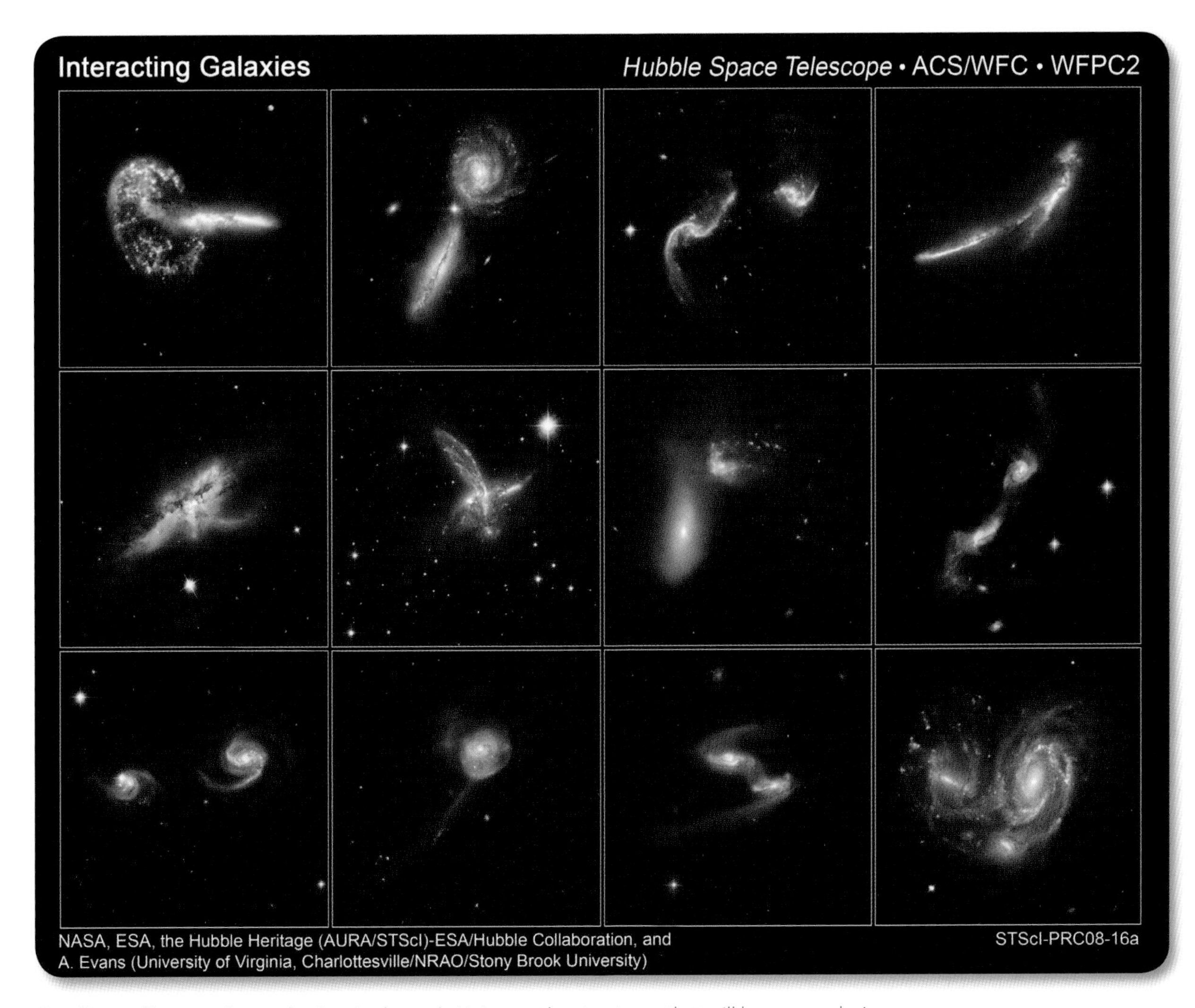

A gallery of interacting galaxies. In the early Universe, the structures that will become galaxies are not much larger than gigantic clusters of stars. These structures are much closer to each other than galaxies are today. Over the next few billion years, these building blocks of galaxies will collide and merge to build the large galaxies that we see nearby. During a collision, gas in the interacting galaxies is compressed and begins to form new stars. Astronomers estimate that only about 1 in 1,000 galaxies in today's Universe is experiencing a collision, but galaxy-galaxy interactions were much more common in the past. NASA, ESA, the Hubble Heritage STScI/AURA-ESA-Hubble Collaboration, and A. Evans (University of Virginia, Charlottesville/NRAO/Stony Brook University)

Since we can actually observe the energy emitted at that time—375,000 years after the beginning—we *know* what the Universe looked like then. Actually, it wasn't yet a very interesting place. There were no stars or galaxies. Instead, the Universe was very simple and almost—but not quite—smooth. There were regions that differed only slightly (1 part in a 100,000) in temperature, and these regions with slightly different temperatures had also slightly different densities. Because stars had not yet formed, there were no sources of energy, and for about 400 million years the entire Universe was experiencing a Dark Age.

As the Universe continued to expand and cool, the regions of higher density—because they could exert a stronger gravitational attraction—began to gather up additional matter from the lower density regions surrounding them. About 400 million years after the big bang, stars and galaxies began to form. We know this because astronomers have now *directly observed* galaxies that formed when the Universe was in its infancy—when it was only about 5 percent as old as it is now.

The earliest galaxies were not at all like the Milky Way. They were much smaller, and there were more of them. Over the next several billion years, these small galaxies merged with each other, building larger galaxies. A collision and merger of two galaxies compresses the gas within them and causes new stars to form. The Milky Way itself has accreted several smaller galaxies. Several billion years from now, it will collide with the Andromeda Galaxy, which is the nearest large spiral, and form an elliptical galaxy.

The Milky Way Galaxy and the Andromeda Galaxy are currently hurtling toward each other at a speed of about 68 miles per second. In about four billion years they will collide, and calculations indicate that they will ultimately merge to form a single, larger galaxy.

This illustration shows a stage in the predicted merger between the Milky Way and Andromeda. In this image, which represents Earth's night sky in 3.75 billion years, Andromeda (left) fills the field of view and begins to distort the Milky Way.

NASA; ESA; Z. Levay

Dark Matter and Dark Energy

So now it seems we know everything—when the Universe began; when galaxies first began to shine; how galaxies have grown over time; how stars form and what makes them shine; and what happens to stars when they finally run out of nuclear fuel.

But wait—nature still has some surprises in store. All of the stars and all of the galaxies and all of the gas and dust in the Universe—that is, everything we can *see*—is only about 4.9 percent of what the Universe is made of. Remember Einstein's equation $E = mc^2$. Matter and energy constitute one hundred percent of the Universe. But 95 percent of that matter and energy is invisible to us. We call what we cannot see *dark matter* and *dark energy*. The very names indicate just how ignorant we are. However, the evidence that dark matter and dark energy do exist is very strong.

Let's look first at the evidence for dark matter. Dark matter does not absorb, reflect, or emit light, but it does exert a gravitational force. The first evidence for its existence came from studies of the rotation of galaxies. Galaxies like our Milky Way rotate. Astronomers expected that the mass of galaxies was mostly concentrated toward their centers where most of the stars are located. They also assumed that stars in the outer regions far from the center, where gravity is weaker, would rotate around the center more slowly than stars in the central regions. This is true in our own solar system. Most of the mass is concentrated in the Sun. Mercury, the planet closest to the Sun, orbits the Sun with a velocity of 47 kilometers per second. Pluto, the most distant (dwarf) planet has an orbital speed that is 10 times smaller: 4.7 kilometers per second.

When astronomers actually measured the orbital speeds of stars in the outer regions of galaxies, they found that the stars were *not* rotating slowly. In fact, the stars were rotating so fast that they would fly off into space unless there was a lot of additional, invisible matter distributed throughout the galaxy. Hence the name "dark matter." This dark matter contributes to the gravitational force acting on the stars far from the center and keeps them from soaring away. Many other observations have confirmed the early results that dark matter must exist.

What is dark matter? Astronomers still don't know. It cannot be ordinary matter that we are familiar with here on Earth—hydrogen, helium, carbon, oxygen, etc.—or we would see it. The most likely possibility is that the dark matter is a new kind of particle with a mass that could be a little smaller than a proton (hydrogen nucleus) up to several times the mass of a proton. There are more than thirty experiments around the world trying to detect such a particle. If it is detected, it will be big news and will likely net a Nobel Prize for the discoverer.

Even if we don't know what dark matter is, it is needed not only to explain the rotation speeds of galaxies but also to explain how galaxies formed so quickly after the big bang. In the early hot Universe, normal matter (hydrogen and helium) interacted with energetic

A supernova in the spiral galaxy M101. The supernova is the bright, bluish star in the upper right portion of the image. It erupted in 2011. Supernovae are extremely bright, as this image shows, and are particularly useful for measuring distances to remote galaxies. Observations of supernovae provided the first evidence that the expansion of the Universe is speeding up. T. A. Rector (University of Alaska Anchorage), H. Schweiker & S. Pakzad (NOAO/AURA/NSF)

The composition of the Universe. Only about 5 percent of all the matter and energy in the Universe is made of the kinds of matter familiar to us here on Earth. Most of that 5 percent is in the form of hydrogen and helium not yet incorporated into stars. The rest—95 percent—is in the form of dark matter and dark energy. Figuring out what they really are is one of the major challenges facing astronomers and physicists. Tim Jones/McDonald Observatory; percentages of dark matter and dark energy updated with the latest results from the Planck satellite by Sidney Wolff

photons and moved so quickly that gravity was not strong enough to cause it to clump together and begin to form stars. Since dark matter does not interact with the photons, it was moving slowly and could begin to clump much sooner. The dark matter thus could build gravitational traps that then captured the ordinary matter and concentrated it so that it became dense enough to begin to build stars and galaxies.

Modern estimates indicate that about 27 percent of everything in the Universe is dark matter. Normal matter contributes another 5 percent. If all of the matter and energy must add up to 100 percent, then we are still missing 68 percent of whatever is out there.

Evidence for what the missing 68 percent might be was first published in 1998. Researchers observed that supernovae, which are stellar explosions that emit an enormous amount of energy, were fainter than they should have been if the Universe were expanding at a constant rate. This is evidence that the expansion of the Universe is accelerating.

Because the Universe is expanding faster now than it was billions of years ago, our motion away from the distant supernovae has speeded up since the light left them, sweeping us farther away. The light then has to travel a longer distance to reach us than if the Universe were expanding at a constant rate. The larger the distance the light has to travel, the fainter the supernovae appear.

In order to accelerate your car, you must step on the gas to supply energy. Energy is also required to speed up the rate at which the Universe is expanding. Since scientists had no idea what the source of this energy might be, they simply called it dark energy. Again, a Nobel prize awaits the person who discovers what causes the acceleration of the Universe.

The Anthropic Principle

Perhaps the most mysterious feature of our Universe is that we are in it. A number of lucky accidents seem to have made our existence possible. Suppose that the tiny density and temperature (10 millionths of a degree) fluctuations in the early Universe had been much smaller. Then calculations show that gravity would not have been strong enough to cause galaxies to form. If the density fluctuations had been much larger, all of the matter might have simply collapsed to form black holes—no stars, no planets, no life.

Suppose the expansion of the Universe were much more rapid. Then the matter in the Universe might have thinned out too rapidly for stars and galaxies to form. If the expansion were much slower, then the Universe might simply have collapsed again—just as a ball falls back to Earth because you cannot throw it upward fast enough to escape the pull of gravity.

Suppose gravity were much stronger than it is. Stellar lifetimes would be measured in years, and there would be no time for life to develop. Even if it did develop, such creatures would have to be very small in order to move around.

There are many other "lucky accidents" without which life as we know it would be impossible. Scientists call these lucky because, even given all that we know about the physical laws that govern the Universe, we cannot explain why gravity is just the way it is, why the expansion rate of the Universe is just the rate that it is, or why the nuclear reactions in stars work the way they do.

As a result, many scientists are beginning to embrace the anthropic principle: that only in a Universe with this apparent fine-tuning will there be intelligent beings who can observe it and wonder why it came about. An extension of this idea is that we may live in a multiverse. This multiverse or mega-Universe could contain many Universes, including our own. The physical laws, including the force of gravity, could be very different in these parallel universes.

It seems unlikely that we will ever know whether we live in a multiverse because there seems to be no way that we could ever observe other universes. Perhaps the concept of parallel universes should be reserved for philosophers and writers of science fiction.

Looking to the Future

Although we have come a long way in understanding how our Universe came to be the way it is, there are many challenges left for future generations of scientists. Are there any forms of life on other planets or moons in our solar system? Are there habitable planets outside our solar system? What is dark matter made of? What is dark energy? New telescopes will continue to be built on the ground and launched into space to try to answer these questions.

Fortunately, astronomers, observatories, and funding agencies like NASA and the National Science Foundation are committed to sharing the latest research with everyone—students and general public alike. Explore the Web, search online image galleries, follow space missions and the building of a new generation of very large telescopes. Share the adventure!

International Dark-Sky Association

The International Dark-Sky Association, based in Tucson, Arizona, is an organization dedicated to combating light pollution and maintaining night skies for current and future generations. More information can be found at http://darksky.org.

Dark Sky Parks of the United States

The association defines its designation of Dark Sky Park in part as "a land possessing an exceptional or distinguished quality of starry nights and a nocturnal environment that is specifically protected for its scientific, natural, educational, cultural heritage, and/or public enjoyment."

Big Bend National Park, TX
www.nps.gov/bibe/

Black Canyon of the Gunnison, CO
www.nps.gov/blca

Blue Ridge Observatory and Star Park, NC
www.mayland.edu/starpark

Canyonlands National Park, UT
http://www.nps.gov/cany/index.htm

Capitol Reef National Park, UT
www.nps.gov/care/index.htm

Chaco Culture National Historical Park, NM
www.nps.gov/chcu/

Cherry Springs State Park, PA
www.dcnr.state.pa.us/stateparks/findapark/cherrysprings/

Clayton Lake State Park, NM
www.emnrd.state.nm.us/SPD/claytonlakestatepark.html

Copper Breaks State Park, TX
http://tpwd.texas.gov/state-parks/copper-breaks

Death Valley National Park, CA
www.nps.gov/deva/

Enchanted Rock State Natural Area, TX
http://tpwd.texas.gov/state-parks/enchanted-rock

Geauga Observatory Park, OH
www.geaugaparkdistrict.org/parks/observatorypark.shtml

Goldendale Observatory State Park, WA
http://www.parks.wa.gov/512/Goldendale-Observatory

Grand Canyon-Parashant National Monument, AZ
www.nps.gov/para/

The Headlands, MI
www.midarkskypark.org/

Hovenweep National Monument, UT & CO
www.nps.gov/hove/

Natural Bridges National Monument, UT
www.nps.gov/nabr/Oracle State Park, AZ
http://azstateparks.com/Parks/ORAC/

Pickett CCC Memorial State Park & Pogue
http://tnstateparks.com/parks/about/pickett

Staunton River State Park, VA
www.dcr.virginia.gov/state-parks/staunton-river.shtml#general_information

Weber County North Fork Park, UT
www.co.weber.ut.us/parks/nfpark.php

Dark Sky Communities in the United States

The IDA defines its designation of Dark Sky Community in part as "a town, city, municipality or other legally organized community that has shown exceptional dedication to the preservation of the night sky through the implementation and enforcement of a quality outdoor lighting ordinance, dark sky education, and citizen support of dark skies."

Beverly Shores, IN
Borrego Springs, CA
Dripping Springs, TX
Flagstaff, AZ
Homer Glen, IL
Sedona, AZ
Thunder Mountain Pootsee Nightsky, AZ
Westcliffe and Silver Cliff, CO

Astronomy Resources

Astronomers try hard to make their latest discoveries and images available to the general public. Some of the opportunities for learning more about astronomy are described in the text. Here are a few more.

On the Web

American Astronomical Society
https://aas.org

Astronomy Now News
http://astronomynow.com/category/news

Astronomy Picture of the Day
http://apod.nasa.gov/apod/astropix.html

Astronomical Society of the Pacific Resource Guides
www.astrosociety.org/education/astronomy-resource-guides

Chandra X-Ray Observatory
http://chandra.si.edu

Infrared Processing and Analysis Center
www.ipac.caltech.edu

Jet Propulsion Laboratory
www.jpl.nasa.gov

National Aeronautics and Space Administration
www.nasa.gov

Sea and Sky Astronomy Resources
www.seasky.org/astronomy/astronomy-resources.html

Sky and Telescope Online Resources
www.skyandtelescope.com/online-resources

Space Telescope Science Institute Office of Public Outreach
http://outreachoffice.stsci.edu

Magazines

Amateur Astronomy Magazine
http://amateurastronomy.com

Astronomy
www.astronomy.com

Astronomy Now
http://astronomynow.com

Sky & Telescope
www.skyandtelescope.com

SkyNews
www.skynews.ca/

Star Parties

Star Party events are too numerous to list. Check with your local science museum, planetarium, university, or college to see if they offer stargazing programs or public lectures.

Glossary

adaptive optics: An optical system with mirrors that can be adjusted rapidly to compensate for distortions of an astronomical image by the atmosphere, thereby producing a sharper image.

annulus: A ring-like structure. One way to think of it is as a circular disk with a circular hole cut in it.

anthropic principle: The idea that physical laws must be the way they are because otherwise we could not be here to observe them.

asteroid: A small, airless, rocky or metallic body that orbits the Sun.

astronomy: Historically, measurement of the positions and motions of stars. Now often used interchangeably with the word astrophysics.

astrophysics: The application of the principles of physics to the study of astronomical objects.

azimuth: The angle along the horizon measured from the north point around to the east. For example, a telescope pointing due south has an azimuth of 180°.

big bang: The rapid expansion that marked the beginning of the Universe.

black hole: A region of space where the force of gravity is so strong that nothing can escape from it, including particles of matter and even light.

dark energy: The mysterious energy that is causing the expansion of the Universe to accelerate.

dark matter: Nonluminous mass, whose presence can be inferred only because of its gravitational influence on luminous matter. The composition of dark matter is not known.

dwarf planet: An object that orbits the Sun and has enough mass to be spherical. It is not a moon of another planet, and it is not large enough to have swept its orbit clean of other objects either by accreting them or by ejecting them to large distances through gravitational interactions.

event horizon: The boundary of a black hole. Nothing inside the black hole, not even light, can be seen by an observer outside the event horizon. Events that occur inside the event horizon cannot affect an observer outside the event horizon.

exoplanet: A planet that orbits a star other than our Sun.

habitable zone: The region around a star in which liquid water could exist on the surface of terrestrial-size planets. Planets in the habitable zone are the ones most likely to have developed some form of life.

hot Jupiter: An exoplanet that has a mass and density similar to Jupiter but that is extremely hot because it orbits very close to its parent sun.

inflation: A short period of rapid expansion that occurred billions of years ago when the Universe was just beginning to expand.

interferometer: An instrument that combines electromagnetic radiation (light, radio waves) from two or more telescopes to obtain images as sharp as could be seen with a single telescope with a diameter equivalent to the separation between the individual telescopes that make up the interferometer.

meniscus: In astronomy, a very thin mirror. A meniscus mirror is not thick enough to preserve its shape when the force of gravity changes as the telescope that holds it moves around the sky. Such mirrors are used to make very large telescopes, but they need an active control system that restores the correct shape required to make sharp images.

nebula: A cloud of interstellar dust and gas.

photoelectric photometry: In astronomy, a technique for measuring the intensity of light from an astronomical source.

planetary nebula: A shell of gas ejected by a hot, low-mass star that is nearing the end of its life. This shell is expanding around the star.

protostar: A very young star still in the process of formation.

quasar: Originally, an object that looks like a star but is at the distance of a galaxy. We now know that the source of energy for a quasar is a black hole in its center that is accreting mass.

reflecting telescope: A telescope in which the largest optical element that collects light is a mirror.

refracting telescope: A telescope in which the largest optical element that collects light is a lens.

resolution: A measure of the smallest details than can be seen in an image.

resolving power: The ability of an imaging device to see details, which are located very close together, as distinct features.

seeing: The blurring and twinkling of stars and other astronomical objects by turbulence in the Earth's atmosphere; a measure of the quality of images as seen through a telescope.

space weather: Processes in space that can affect the Earth and its nearby environment. An example is the ejection of matter by giant storms on the Sun, which can cause auroras on Earth.

spectroscopy: In astronomy, the analysis of the various wavelengths of electromagnetic radiation, including light, emitted by astronomical objects. Spectroscopy can be used to determine chemical composition, temperature, mass, distance, and other properties of astronomical objects.

supernova: Created by the death of a star, this is a spectacular stellar explosion that can briefly outshine an entire galaxy.

transient event: A sudden and brief appearance of a source of light in the sky. Some events last only a few seconds; others can be observed for days, weeks, or even a few years.

transients: Objects that appear suddenly but were not visible in earlier observations. A supernova is an example of a transient.

transit: The movement of a small object in front of a larger one. An example is the transit of a planet in front of a star.

turbulence: Random motions of gas. Turbulence in the Earth's atmosphere causes stars to twinkle and makes images seen through telescopes look blurred.

Index

About the Author

Emily Acosta/LSST Corporation

After receiving her PhD from the University of California, Berkeley, Dr. Wolff moved to Hawaii, where she was involved with the astronomical development of Mauna Kea, which is now recognized as the best observing site in the northern hemisphere. In 1984 she became the Director of Kitt Peak National Observatory, and in 1987 was named Director of the National Optical Astronomy Observatory, where she led the design and development phases for the twin 8-meter Gemini Telescopes and the 4-meter Southern Observatory for Astronomical Research. Most recently, she led the design and development phase of the 8.4-meter Large Synoptic Survey Telescope, which is designed to survey the entire observable sky every few nights and to conduct a 10-year search for objects that change in brightness (e.g., stellar explosions) or change position (e.g., potentially hazardous asteroids that might crash into the Earth). Gemini South, LSST, and SOAR are all located on Cerro Pachón in Chile. A view point that looks toward Cerro Pachón has been named Mirador Sidney Wolff in recognition of her leadership role in developing these facilities. Dr. Wolff has published over ninety scientific papers on star formation and stellar atmospheres in refereed journals. She has served as president of the two leading professional astronomy societies in the United States—the American Astronomical Society and the Astronomical Society of the Pacific. Dr. Wolff is also the coauthor of several textbooks widely used in introductory college astronomy courses.